Rick Hattenhoff

# Minerals

## AND HOW TO STUDY THEM

**By the late JAMES D. DANA**

SYSTEM OF MINERALOGY. *Seventh Edition.*

Rewritten and enlarged by the late Charles Palache, the late Harry Berman, and Clifford Frondel.

Vol. I.   1944.
Vol. II.   1951.
Vol. III.   **1962**

MANUAL OF MINERALOGY. *Seventeenth Edition.*

Revised by Cornelius S. Hurlbut, Jr.


**By the late EDWARD S. DANA**

A TEXTBOOK OF MINERALOGY. *Fourth Edition.*

Revised by the late William E. Ford.   1932.

MINERALS AND HOW TO STUDY THEM. *Third Edition.*

Revised by Cornelius S. Hurlbut, Jr.   1949.

# Minerals

## AND HOW TO STUDY THEM

BY

The Late EDWARD SALISBURY DANA

REVISED BY

CORNELIUS S. HURLBUT, Jr.

*Third Edition*

John Wiley & Sons, Inc.   New York • London • Sydney

# Preface

When this book appeared in 1895, it was the first of its kind to present facts about minerals in a manner which was interesting to the amateur. It was widely accepted then and continued in popularity through half a century even though no major revision was made during that time. With the ever increasing interest in mineralogy, it seemed desirable to bring the book up to date.

Although much of the book has been rewritten, an effort has been made to maintain the same point of view for the same reader, the beginner in mineralogy. The sections on *Crystallography* and *Physical Mineralogy* cover much the same ground as previously, but the chapters on *Chemical Properties of Minerals* and *The Use of the Blowpipe* have been somewhat shortened while still retaining the essentials necessary for the beginner. The chapter on *The Description of Mineral Species* has been completely rearranged and rewritten. The minerals are arranged according to the almost universally accepted chemical classification, and the description of each species follows a common pattern to make the pertinent data more accessible. In the original edition the minerals that are compounds of the same metallic element were grouped together. Much thought was given to the type of classification before making the change, but it was felt desirable for the student to begin his study of minerals with the same classification he would most surely use if he were to go on to advanced work.

The author wishes to acknowledge the aid given by Dr. Howard T. Evans, Jr., in writing the pages on *Crystal Growing*. The color photographs were taken by Mr. Carl Miller of Ward's Natural Science Establishment.

HARVARD UNIVERSITY                                              C. S. HURLBUT, JR.
*Cambridge, Massachusetts*

# CHAPTER I

# Minerals
# and Mineralogy

We are to learn about minerals and how to study them, but, before we begin, we must understand clearly what substances may be called minerals and what specimens rightfully belong in the collection that everyone who wishes to become a mineralogist must make.

It is essential at the outset that we understand that minerals are the materials out of which the earth is built. Although we live and move upon the earth, we know little about it by actual contact or direct observation. It is possible, however, to measure its size and shape, to determine its density as a whole, and to study its surface features and the changes they have undergone; but of the materials of which it is made we can know little beyond those that form the surface upon which we walk. The miner digs down a little distance, the oil-well driller reaches down still deeper, and we may examine the materials that their work brings up. Perhaps we can go down with the miner and see the rocks in place, but the deepest mines reach to depths of less than two miles. Although this seems very deep to one who is let down a shaft in a cage, it is only a little way compared to 4,000 miles, the distance to the center of the earth. Even the deepest oil well reaches down to a depth of only a little over three miles.

Our knowledge, to be sure, is increased a little by the fact that we find now on the surface of the earth rocks made, we have reason to believe, of materials brought up in a molten condition from great depths below. Igneous rocks such as form the Palisades along the Hudson River as well as the lava thrown out by volcanoes

1

had their origin many miles below the present surface of the earth. A careful study of these rocks gives us some idea of the kinds of matter and the conditions under which they exist in depth. Furthermore, the fact that the astronomer has weighed the entire earth and found its density to be nearly twice as great as that of the rocks on the surface suggests the heavy nature of the mineral material that must make up the interior.

Thus the mineralogist is limited to the study of the upper part of the earth's crust which he can reach with his hammer. He cannot extend his collection much beyond this unless, indeed, he includes some of those rare visitors from outer space—called *meteorites*—which occasionally fall to the earth.

What do we learn from a study of this hard rocky material of which the earth, so far as we can examine it, is made? We find, in the first place, that it consists of different kinds of rocks, a few of which have familiar names, such as granite, marble, sand, sandstone, trap, and slate. On closer examination we find that each rock consists of different substances, each having certain peculiarities or properties of its own by which it can be recognized. It is to these individual substances that the name *mineral* is given.

Thus, more particularly, the sand of the seashore can be separated without much difficulty into various kinds of grains. All the grains of each kind are chemically alike with certain properties of hardness, density, luster, and color, which enable us, after a little practice, to distinguish the different kinds with comparative ease.

Most of the grains are of one kind, clear and glassy, hard enough to scratch glass, and, as we learn to know them better, we call them *quartz*. There are also black grains; some of these are heavy and jump to a magnet, and often they are sorted out by the waves into little rifts on the white sand; these are called *magnetite*. There are other black grains, too, which the magnet does not attract; these may be *ilmenite*. Perhaps some are red and glassy and are fragments of *garnets*. There may be still others, depending upon where the sand comes from and what kind of rock has been ground up by nature's mill and sorted out by the water to make the sand.

If a piece of granite is examined closely, here too it is possible to distinguish several kinds of minerals, though not quite so easily.

There are hard glassy grains with irregular surface which, like the greater part of the sand grains, are *quartz*. There are white or yellow or pale flesh-red fragments, also hard, though not so hard as the quartz, but which show one or two smooth surfaces of fracture; these are *feldspar*. Then there is the *mica*, more easily recognized, which is either white and silvery, or shiny black (in some granites both kinds are found), and which with a touch of the knife blade separates into very thin scales or leaves. Besides these there may be a little coal-black *tourmaline*, some bright red *garnets*, and smaller amounts of other kinds which we shall learn later. If a cavity or open space can be found in the granite, it is often possible to find in it the same minerals, only in larger and more distinct form and commonly in regular shapes called crystals.

If, instead of a coarse-grained rock like granite, we examine one that is fine and compact such as the traprock of the Palisades on the Hudson or a lava flow from New Mexico, it will probably appear to be quite uniform to the eye. However, if we crush some of it to powder, the magnet will pick out some magnetite, as from the seashore sand. Or the skillful mineralogist may make a slice thin enough to be transparent, so that he can study it under the microscope and then recognize a variety of different minerals. In seams and cavities in these rocks other minerals are often found which differ from those in the massive rock.

In places we find a rock like the white marble of Vermont which examination shows is made entirely of the same chemical substance, and which has throughout the same properties of hardness, density, and color. It is a mineral itself, *calcite*, and is called a *monomineralic* rock, unlike most rocks, which are a mixture of different minerals.

These different kinds of substances, then, which make up the rocky material of the earth's crust, and into which we can separate the seashore sand, the granite, and most other rocks, are called *minerals*. Each one is, first of all, a chemical element or compound. Moreover, if well crystallized, each has, in addition, a shape of its own by which it may be distinguished. It has also certain properties of hardness, density, luster, color, and transparency. Because to it belong all these different properties, which distinguish it from other kinds, it is called a *mineral species*.

It is the work of the mineralogist to study all these different minerals and to learn the properties of each; how they are classified and distinguished from one another; how they occur in nature, and something about their practical uses.

All the knowledge that the many mineralogists have gained after long years of patient observation and study, both in the field and in the laboratory, has been arranged in systematic form and makes up the *science of mineralogy*.

Since mineralogy includes not only the description of minerals, but also the way in which they occur, the minerals with which they are associated, and ultimately their origin, it bears a definite relation to the broader science of *geology*. Geology treats of the history of the earth and all the changes through which it has gone as read in the record of the rocks. It is essential, therefore, that the geologist know something about mineralogy; and at the same time it is desirable for the mineralogist to know something about geology.

There is also a profitable exchange of knowledge between the chemist and the mineralogist. With the exception of the gases of the atmosphere, minerals are the source of all the chemical elements; and the chemist learns from the mineralogist in what minerals, at what localities, and in what amounts they are to be found. A chemical classification of minerals is considered by most professional mineralogists to be the best. It is from the chemist that the mineralogist learns not only how to arrange his minerals in chemical groups but also how to make chemical analyses to determine one of the most fundamental properties of minerals— the chemical composition.

It is not to be inferred that, before one should undertake a study of minerals, he must first be a geologist and a chemist, although, for advanced work in mineralogy, a familiarity with these related sciences is almost essential. For the beginner the reverse may be true, and he will discover after a short while that he is learning some geology and chemistry as a by-product of his study of minerals.

On the foregoing pages some of the properties of minerals have been pointed out in an effort to build a picture of a mineral in general terms. Before proceeding further we should formulate a more specific definition. The word *mineral* has today several different meanings, but to the mineralogist it is *any naturally occurring chemical element or compound formed as a product of inorganic processes*. The earlier definition of mineral was not so restricted, for anything that was dug from a mine was called a mineral. Most mines produce not single substances but mixtures of many things and vastly different things at different localities. The term mineral has gradually taken on more specific meaning until at present the mineralogist uses it only for those substances that answer the above definition.

Let us consider the implications of the definition. We find that a mineral must occur in nature; anything made by man, no matter what its physical and chemical properties, cannot be a mineral. Thus the many beautiful kinds of salts made by the chemist are not minerals. The rock salt or sodium chloride which is mined in fine clear cubical blocks is the same sodium chloride which, as the table salt of everyday life, is so commonly used. But the table salt obtained from evaporating sea water or the brines of salt wells, or from the solution of crude rock salt, though in crystals as fine as or finer than those found in the rocks, is not a mineral because it was not made by nature alone. So, too, the fine crystals of copper sulfate (see p. 64) made in the laboratory do not find a place in a mineral cabinet, though the much less perfect specimens of the same material found in some of the Arizona mines do.

It must be pointed out, moreover, that many specimens made in the laboratory are very minute and much less beautiful than those of nature. The chemist in the laboratory has only a limited time for his experiments, whereas nature works gently with unlimited time.

However, in recent years many attempts have been made in the laboratory to imitate the processes of nature and produce "artificial minerals." As a result, many minerals have synthetic counterparts that rival or even excel the natural substances in perfection of development. An outstanding example is the manufacture of rubies and sapphires that, when cut into gemstones, are difficult if not impossible to distinguish from natural stones.

The restriction that a mineral must be a chemical element or compound means that its composition can be expressed by a chemical formula. Thus all mixtures, even if they are quite uniform and homogeneous, are eliminated. In the past, mineral names have been given to certain apparently homogeneous substances which have been shown by more recent work to be fine-grained mixtures or aggregates.

Finally, a mineral must be formed as a product of inorganic

processes. Any substance, therefore, resulting directly from plant or animal life, even if it closely resembles natural inorganic materials, cannot be called a mineral. For this reason, the mineralogist usually excludes from his collections many substances such as the pearl of the oyster and the shell itself, the lime of the bones of animals, and the opal-like form of silica secreted by the growth of plants, such as the tabasheer found in the joints of the bamboo. In general, such substances formed immediately by the processes of animal or vegetable life are not called minerals.

Unfortunately, the term mineral is not used in the restricted sense by everyone; when the economist speaks of the *mineral resources* or the *mineral wealth* of a country, he refers not only to the minerals of the mineralogist but also to coal and petroleum, which are organic in origin.

# CHAPTER II

# *Some Preliminary Hints*
# *on How to Study Minerals*

We have seen that a mineral has certain properties of form, hardness, density, luster, and color which collectively characterize it and frequently enable us to separate it from other minerals. These are known as its *physical properties.* The most important property of a mineral, however, is its chemical composition, and dependent directly upon it are its *chemical properties.* These two sets of properties will be described in some detail in subsequent chapters; but first it is necessary to describe briefly how to study minerals, if we wish to learn as much as possible about each species with the least effort.

The student must first of all use his eyes and other unaided senses in studying minerals; in other words, he must gain all the information he can about minerals by looking at them and handling them. If he learns to do this wisely, he will be surprised to find how keen his senses become and how much he can tell by merely inspecting his specimens. However, as he gains experience, he will see that this method carries him only to a certain point, and he recognizes the importance of confirming his first conclusions by more positive tests. The appearance of specimens of even the common species may, if one depends upon it alone, lead one quite astray. The old saying, "all that glitters is not gold," and the names "fool's gold" and "false galena," express the conclusion that the senses unassisted may readily be deceived.

The trained eye of the mineralogist will show him at a glance whether a mineral has the regular geometrical shape of a crystal. It will show him also whether the mineral has the natural, easy, smooth fracture of many crystalline substances, called *cleavage*, or only a rough irregular fracture. It will tell him the color, and whether it is transparent or opaque, as well as the peculiarities of luster that may be characteristic of a given mineral.

The touch will indicate whether the "feel" is greasy, as is true of talc and a few other very soft minerals. When one "hefts" a mineral, one may recognize at once that it is heavy or light as compared to familiar substances of the same appearance. The common minerals, quartz, feldspar, and calcite, have nearly the same density, and one can easily become so accustomed to them that a piece of gypsum seems light and a piece of barite seems heavy. Similarly, a piece of aluminum or, more particularly, magnesium seems very light because it is instinctively compared with apparently similar but much denser metals which we are more accustomed to handle.

The *taste* in some minerals is a distinguishing characteristic; the *odor* is occasionally a useful property, such as the clayey odor given off by some minerals when they are breathed upon. It does, however, require some study and experience before the senses are so alert that all the properties to be noted are perceived at once and rightly evaluated. To this end everyone should strive, for one of the great benefits to be derived from the study of mineralogy is that it cultivates and stimulates the powers of observation.

When the senses have gleaned all that is possible, simple tests to aid them should be used. Touching the smooth surface of a mineral with the point of a knife serves to show whether it is relatively soft or hard. The color of the powder obtained by rubbing a mineral across a plate of unglazed porcelain, or scratching it with a knife, is called the streak; if the streak is quite different from the color of the surface, as it is in some minerals, it constitutes a very important property.

If inspection of the specimen or the tests already mentioned fail to identify a mineral, more careful tests must be made. These include the density or specific gravity; the use of the blowpipe; and a number of simple chemical tests, to show the presence of certain elements. After these are made there are still others, which include the refined methods of the trained mineralogist with his precision goniometer for measuring crystal angles, the polarizing microscope for the study of optical properties, x-ray analysis for the study of the internal structure, and the accurate quantitative chemical analysis. With these methods and others that are constantly being developed, most of nature's secrets may be learned and the properties of each mineral may be thoroughly studied.

## SUGGESTIONS ON MAKING A MINERAL COLLECTION

A very important part of the study of minerals is the student's own collection, for everyone who wishes to learn mineralogy must have a collection of his own to examine and experiment upon. It is very desirable that the school or college have a larger cabinet for reference and study, but such a cabinet does not take the place of the individual collection, which will be studied, arranged, labeled, and handled over and over again until every specimen is perfectly familiar.

Furthermore the student should obtain his specimens as far as he can by collecting for himself, and thus see the minerals in their natural surroundings. Even if he lives in a region that does not seem at first to afford many, he can find something that is worth keeping until he obtains better. Occasionally he will have the opportunity to make trips to some of the noted localities, where he can find a great variety of minerals. There is nothing more delightfully instructive and refreshing than to spend a day in the open air, with a good hammer in hand, a bag for the specimens, and plenty of soft paper to wrap them in.

The hammer should be of hard steel that will not chip on the edges; it may weigh a pound to a pound and a half, the face should be square or slightly oblong, and the edges should be sharp, with the back in the form of a wedge, as seen in Fig. 1 (one-fourth natural size). A cold chisel or two, for working into cracks or crevices, will often be found useful. Also a small light hammer with a sharp edge for trimming specimens may prove valuable, for a blow from a heavy hammer will often shatter a specimen.

It is better as a rule not to break the crystals out of the rock. A detached crystal of garnet is interesting when quite perfect, but in general the crystal is most interesting and instructive when in its own surroundings. The seller of minerals soon discovers this, and it is unfortunately not an uncommon trick at some localities for the local collector working for his daily bread to exercise his ingenuity by mounting a loose crystal in a mass of rock in which it never belonged, thus to increase the value of the specimen and deceive the unwary purchaser.

The student is not advised to spend a great deal of money on buying specimens, particularly at any one time, for today many mineral localities are accessible by automobile from almost any point in the United States, where it is possible to see minerals in place and to collect fine specimens of many minerals.

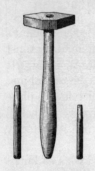

However, a little money is by no means thrown away if judiciously expended from time to time, for it will serve to buy a few small characteristic specimens of the common species and pure fragments for blowpipe tests. Fine specimens, especially of the rarer species, are now very expensive. Fortunately, sufficiently good specimens of the minerals that it is important for the student to know well may be obtained for very little money.

Fig. 1.   Hammer and Cold Chisels.

It is better to collect, as far as possible, small specimens rather than large, such as will fit in a little paper tray 2 inches square, or 2 by 3 inches, or at most 3 by 3 inches. These trays are inexpensive and are very useful for the arrangement and preservation of a collection. If the specimens are placed loose in a drawer, it can be opened only a few times without throwing them into confusion and separating them from their labels; sooner or later they will be badly injured. A depth of half an inch is sufficient for the tray, but the drawers, if possible, should not be less than $2\frac{1}{2}$ or 3 inches deep. All the specimens in a collection should be carefully labeled, particularly as regards *locality*.

Even if considerable care is exercised, specimens will become separated from their labels. In order to prevent permanent separation, it is well to paint a number directly on the specimen as soon as it is acquired. The number with the locality and other pertinent data should be entered in a catalogue to which reference can be made. A good way to insure permanence of the number is to paint on the mineral in an inconspicuous place a small area of white enamel and, when this is dry, to write the number on it in India ink.

Large, well-crystallized mineral specimens are usually not available to the average collector because of their scarcity, and such specimens of many minerals can be seen only in museums. There is a rule, however, to which there are few exceptions, that the

FIG. 2. Binocular Microscope and Micromount.

FIG. 3. Magnified Micromount.

smaller the crystal the better it is formed and the more faces it possesses. Because of this fact, some collectors search for small, well-crystallized specimens that can be seen well only through a microscope.    These are mounted individually on pedestals which in turn are cemented to the bottom of small boxes or trays.    Figure 2 shows a *micromount* in position to be observed with a microscope. Figure 3 is a picture of the same micromount magnified twenty times.    When viewed in this manner, these tiny specimens rival or even surpass in beauty and perfection of development larger specimens of the same minerals.

# CHAPTER III

# Crystals
## and Crystal Habits

The principal properties of minerals, by which one species is distinguished from another, have been briefly alluded to in the preceding chapter. It is now necessary to study some of these properties more fully.

First the *physical properties* will be considered. These include crystal form and habit, cleavage, fracture, hardness, tenacity, elasticity, density, color, luster, degree of transparency, and a few others. The present chapter is limited to a discussion of crystals, their forms, habits, and aggregates.

## THE GENERAL NATURE OF CRYSTALS

If we examine the specimens of the different mineral species in a good collection, we see that many of them are regular solids with smooth faces which, as we study the subject further, we find to be characteristic of each individual species. These regular forms are called *crystals*. The cubes of fluorite (Fig. 4) or galena, the six-sided prisms of quartz (Fig. 5), the twelve-sided, twenty-four-sided, or even more complex forms of garnets are common examples of crystals. Although these minerals may show other crystal faces, it is their "habit" to crystallize in the forms indicated. Thus *crystal habit* means the common and characteristic form assumed by a mineral including the general shape and irregularities of growth. Even when a specimen does not show a regular external form, there is usually a definite crystalline structure which may be shown in the easy, smooth fracture

13

called *cleavage*, such as that of calcite or fluorite. How is this regularity of form to be explained?

The physicist, as the result of his studies of the structure of different kinds of matter, has concluded that everything is built up of minute particles called atoms, which are much too small to be

Fig. 4.    Fluorite, Group of Cubic      Fig. 5.    Quartz Crystal.
Crystals.

seen even by the strongest microscope. In a solid body, as a lump of iron, ice, or sulfur, he thinks of these atoms grouped into molecules and bonded together by a strong force of attraction called cohesion, so that it requires a hard blow or great pressure to change its shape. In a liquid body he thinks of molecules as free to move or roll over each other, so that the liquid takes at once the shape of the vessel in which it is contained, whatever that may be. In a gas, he believes that the molecules are separated from each other by a long distance compared with their size, and that they are darting about very rapidly, colliding with each other and any confining surface. Consequently the gas at once fills entirely a vessel into which it is introduced and presses against its sides. The pressure of the external air, for example, is shown by the collapse of the cheeks when the air within the mouth is drawn away.

The relation between these molecules explains the condition of a substance as solid, liquid, or gaseous: for example, the distinction between ice, water, and steam.

When a liquid turns into a solid because the temperature falls, as when water freezes, or liquid sulfur or molten iron hardens on cooling, the force of cohesion comes into play to bind the particles together into a rigid mass. Likewise, when by slow evaporation from a solution, as of salt or alum in water, the dissolving liquid is removed, the substance in solution passes back into the solid form under the action of this same force of cohesion. Thus the solid is formed from the liquid by the action of the forces acting between these little particles. Further, if all the atoms are of one kind, as in a given chemical element, and if there are no hindering causes, these atoms build themselves up according to some regular

FIG. 6.   Snow Crystals.

pattern, and the external result is the geometrical form that is called a crystal. It is somewhat as if the atoms were little building-stones, built up into a solid structure by forces acting between them and causing them to arrange themselves after a definite manner when they are free to do so.

This regular building of the atoms not only takes place as has just been shown when a solid is formed from a liquid but also when a solid is formed directly from a gas. Water vapor in the air, if cooled sufficiently, is arranged in the form of a solid snowflake. The little snow crystals that fall silently through the atmosphere and that we catch on our coat sleeves on a cold winter day have wonderful regularity and beauty. Figure 6 shows some of the many forms of snow crystals.

It is not always easy to make good crystals, whether we start from a liquid or from a gas, partly because we cannot give the time required for the perfect process, partly because there are other hindering conditions. Some crystals, however, can be grown in a limited time (as described on p. 63); the growth of an octahedral crystal of alum can be watched from day to day, and a large and fine crystal may be the reward of skill and patience.

In nature's laboratory the conditions are more favorable, particularly because there is never any time limit, and the many beautiful and complex crystals of minerals with brilliant faces show the result. Even here, however, the building process often cannot go on freely, and imperfect crystals, or perhaps a mass of only a confused crystalline aggregate without distinct external form, may be all that is produced.

The quartz, the feldspar, and the mica in the rock called granite have usually formed in such a way that they interfere with each other, and none of the minerals has had an opportunity to build itself up into perfect crystals. In spite of this the student who understands the optical study of thin sections of a rock in polarized light can prove that each grain, formless though it may be externally, is *crystalline* and has the internal atomic structure of a crystal. In a cavity in the granite we are not surprised to find crystals of quartz and feldspar, perhaps also of mica, as the existence of the cavity means that each mineral has had an opportunity to exercise its tendency to build itself regularly with some of the freedom that a perfect crystal requires.

Another familiar example of crystallization is the ice covering a pond, which is as truly crystalline in structure as the perfect snow crystal; but here there are no individual crystals. The slow dissection of the mass, however, under the melting action of the sun reveals something of the regularity in the building, and the same regularity is proved by an examination in polarized light. Often in the freezing of a little pool of water on a sidewalk the formation of the slender crystalline ribs of ice may be watched as they shoot out, forming a framework which may soon lose its distinctness as the entire surface is frozen.

We learn, then, that a crystal is the geometrical form assumed by a chemical substance, if it is free to form without interference when it passes into the solid state from that of either a liquid or a gas. The crystal is, therefore, the *outward* expression of the orderly internal arrangement of atoms. For this reason, crystal form is the most important of all the physical properties of a given species and the one that in general most definitely characterizes it. Since it is possible for us to grow crystals in the laboratory, it must be pointed out that not all crystals are minerals but only those that nature alone has produced.

Most minerals, even though no crystals are visible, are crystalline, but a few may be *amorphous*. A crystalline substance is one in which there is an orderly internal atomic arrangement even if there is no outward sign of it, whereas an amorphous substance lacks this orderly structure. The difference is somewhat like that between a pyramid built of bricks laid in regular rows and one in which the bricks are tumbled in without order. Thus a piece of clear quartz is said to be crystalline, and a piece of glass which the eye alone could not distinguish from the quartz is amorphous. If the quartz were fused and allowed to cool, a quartz glass would result with exactly the same chemical composition but the atoms in the glass would not have arranged themselves into the orderly array of the crystal and would thus be amorphous.

How then are we to tell whether or not a massive mineral is crystalline? Cleavage, as explained later, is, like crystal faces, an expression of the orderly atomic arrangement. If cleavage is present, we can say with certainty that the mineral is crystalline. If no cleavage is present, an examination in polarized light will help in making a decision. For example, the bright colors seen in a thin fragment of a quartz crystal in polarized light show at once, to one who understands the subjects of optics, that it is crystalline. The fused quartz on the other hand will show no such colors. If the mineral is opaque, it may be necessary to take an x-ray photograph. A crystalline substance will give a regular pattern recorded on a photographic film, whereas an amorphous substance will give no pattern.

Another method, known as etching, aids the skillful mineralogist in determining an ordered atomic structure. He allows some liquid (or gas) that has the power of dissolving the substance to act upon a smooth surface for a short time. Then it is removed, and the surface is cleaned and examined under the microscope. Often a multitude of little cavities or pits are found on the surface, the shape of which shows clearly how the structure is built up. The process of etching is as if the stones of the pyramid spoken of were so smooth and closely fitted that no joints were visible, and the mason should pull out a number of bricks till he could see the pattern. Figure 7 shows the figures etched by hydrofluoric acid on the faces of a crystal of quartz; the variations reveal the complex structure of this mineral.

In general it can be said that although many minerals are massive they are usually crystalline and the amorphous state is the ex-

ception.  There are, however, certain minerals or varieties of
minerals that seem to be intermediate between the distinctly
crystalline and the amorphous states; these are called *cryptocrystalline*.  Many fine-grained varieties of quartz are cryptocrystalline and resemble amorphous substances.  When these
minerals are viewed under a high-powered polarizing microscope,
one can see they are an aggregate of myriad tiny crystalline quartz
particles.  Moreover, an x-ray photograph
would give the same pattern as that given by
a single large quartz crystal ground to a fine
powder.

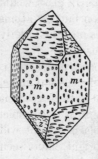

As already mentioned, a small crystal is just as
perfect and complete an individual as a similar
one of great size.  Among the crystals of a
given species there is no relation between size
and age, as there is among the individuals of a
species in the animal and the vegetable king-
doms.  Some crystals are so minute as to be

Fig. 7.     Etched
Quartz  Crystal.

microscopic; others may be of enormous size,
as the large beryl crystals from Maine and the
even greater spodumene crystals in the Black
Hills of South Dakota, some of which measure over 40 feet in
length.  A cave opened many years ago at Macomb, New York,
contained 15 tons of great cubic crystals of fluorite; another cave
in Wayne County, Utah, contained a large number of enormous
crystals of gypsum, some of them 3 feet or more in length.  But
the very small crystals and the like ones of enormous size are not
essentially different except in the comparatively unimportant
respect of magnitude.

Figure 8 is a photograph of a group of gypsum crystals from
Naica, Mexico, the largest of which is over 4 feet long.

Nevertheless, there are many interesting points of resemblance
between crystals and living plants.  Crystals, as well as plants,
*grow* and, under favorable conditions, so rapidly that the increase
in size may be watched not only from day to day, but from hour
to hour, or even from minute to minute.  The complex forms that
are built up especially in such cases of rapid growth are often
wonderfully plantlike in aspect.  As everyone has noticed the
delicate frost figures form quickly upon a windowpane in winter;
other, more permanent examples are the arborescent or dendritic

forms of native gold, silver, or copper. Indeed the terms used in describing them are given because of their resemblance to forms of vegetation. Furthermore, as a wounded plant tends to heal itself when, for example, a branch has been broken, so, too, a broken crystal may be more or less healed, but with crystals the material that repairs the injury must be supplied from an outside source.

Fig. 8. Gypsum Crystals, Naica, Mexico.

Thus the silica to mend a broken quartz crystal must come from a foreign solution, and the crystal itself only directs the way in which the atoms from the solution are laid down. It is interesting, however, that the growth takes place more readily on a surface of fracture than on a natural crystal face. In this way the grains of quartz in a sandstone, formless because they are only rounded fragments, often tend to build themselves into complete crystals.

Although a crystal never has an old age in the sense that a plant or an animal does, nevertheless many crystals change as time goes on, if subjected, for example, to the corroding effects of some foreign solution.

Even beautiful gems, such as the sapphire, emerald, and topaz, which are hard and comparatively insoluble, have this tendency to

undergo what is called chemical alteration, with the loss of their
beauty and a change of chemical substance. This alteration is
spoken of again in a later part of the chapter, where pseudomorphs
are described, but it is worth noting here because it is somewhat
analogous to the change that old age brings to a living organism.

### CRYSTAL SYMMETRY

If one is privileged to handle well-formed crystals of a variety
of minerals, he will be quick to see that there is a wide difference
in them. The size, shape, appearance, and number of faces differ

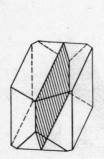 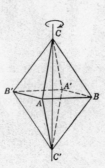 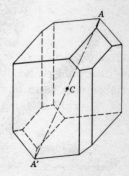

Fig. 9.  Symmetry      Fig. 10.  Symmetry      Fig.  11.  Symmetry
Plane.                 Axis.                   Center.

from crystal to crystal. On one all the faces may be similar,
whereas on another only two similar faces may be found, and these
may be located on opposite sides of the crystal. A little study
will show that like faces are arranged symmetrically on the crystal
and that this symmetry is the same for all crystals of any given
mineral species.

Three types of symmetry are to be found in crystals, known as
the elements of symmetry; they are: (1) symmetry across a
plane; (2) symmetry about an axis; (3) symmetry about a point.

A *plane of symmetry* is an imaginary plane passed through a
crystal dividing it in half, so that each half is the mirror image of
the other. If it were possible then to split a crystal in half along
this plane and place one half against a mirror, the image would
appear to restore the other half. Figure 9 illustrates a plane of
symmetry. Some crystals have as many as nine planes of symmetry although some have none.

An *axis of symmetry* is an imaginary line through a crystal about which the crystal may be revolved and may repeat itself two or more times in a complete revolution. There are 2-, 3-, 4-, and 6-fold symmetry axes. Figure 10 illustrates a 4-fold axis.

A crystal is said to have a *center of symmetry* if an imaginary straight line can be passed from any point on its surface through its center to a similar point on the opposite side. Both Figs. 9 and 10 have symmetry centers as well as axes and planes; Fig. 11 has only a center of symmetry.

It is possible for certain crystals to lack all elements of symmetry, but only one or two such minerals are known. One might think that if these three symmetry elements were combined in different ways, a hopelessly large number of combinations would be possible. Actually there are only 32 such combinations, and they are called the *32 crystal classes*. To the beginner in mineralogy even this number may seem large, but he should not be discouraged for only 10 or 12 of the 32 classes are represented in the common minerals.

## CRYSTAL SYSTEMS

Although each of the 32 crystal classes mentioned above differs in symmetry from the others, certain of them can be grouped because of similarities. There are 6 larger groups, the *crystal systems*, named as follows: I, isometric; II, tetragonal; III, hexagonal; IV, orthorhombic; V, monoclinic; VI, triclinic. All the crystals in a crystal system are referred to the same lines of reference, the *crystallographic axes*. These axes will be described in the discussion of the crystal systems on the following pages.

## Isometric System

All the crystals of the isometric system can be referred to three equal axes at right angles to each other. These imaginary lines intersecting at the center of the crystal are the *crystallographic axes* (Fig. 12), and it will be seen that the arrangement of faces about the six ends of the axes of a given crystal is always the same.

Of the 32 crystal classes, 5 are in the isometric system; and, although they have different symmetry, they can still be referred to these same crystallographic axes. Only 3 of these 5 classes are common enough in minerals to deserve our attention.

The most common of these classes is the *hexoctahedral* or the *galena class.* It has the highest crystal symmetry with the following symmetry elements: nine planes, four 3-fold axes, three 4-fold axes, six 2-fold axes, and a center. The principal forms* in this

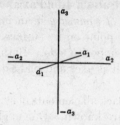

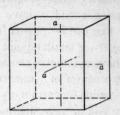

FIG. 12. Isometric Crystal Axes.      FIG. 13. Cube.

class are the cube, octahedron, dodecahedron, trapezohedron, tetrahexahedron, and hexoctahedron. In the following descriptions, the perfect geometrical solid is described, but it should be

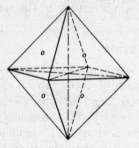

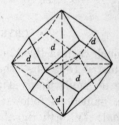

FIG. 14. Octahedron.      FIG. 15. Dodecahedron.

made clear at the start that most crystals are not model perfect and are frequently malformed as described on page 46.

**Cube.** The cube has six equal faces, each one of which is a square, and the angle between any two faces is a right angle, or 90°. It is shown in Fig. 13. Galena and fluorite often occur in cubes.

**Octahedron.** A regular octahedron (Fig. 14) has eight like faces, each a triangle with equal sides and equal (60°) angles; the

* The term *form* is used here in the special sense of the crystallographer. Although it may be used to express the general outward configuration of the crystal, it is usually reserved to indicate all those crystal faces that have like positions with respect to the crystallographic axes and the elements of symmetry. Thus the cube is a crystal form composed of six faces, the octahedron, eight faces, etc.

angle between any two adjacent faces is 109° 28'. Magnetite is often in octahedrons.

**Dodecahedron.** The rhombic dodecahedron (Fig. 15) has twelve equal faces, each of which is a rhomb with plane angles of

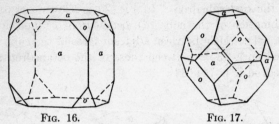

FIG. 16.                    FIG. 17.

Cube and Octahedron.

60° and 120°, the angle between two adjacent faces being 120°. This is a common form of garnet.

All these three forms may occur on the same crystal. Thus crystals of galena often show the cube and octahedron. Figure 16 is generally described as a cube modified by an octahedron, and

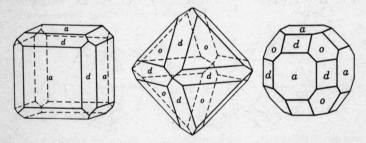

FIG. 18.   Cube and    FIG. 19. Octahedron and   FIG. 20.    Cube,
Dodecahedron.          Dodecahedron.            Octahedron, and
Dodecahedron.

Fig. 17 as an octahedron modified by the cube. If a cube is cut out of a block and the solid angles are sliced away carefully, the new surfaces making equal angles with three cube faces, the result is an octahedron. It is seen that the octahedral faces are little triangles on the solid angles of the cube and are equally inclined to the three cube faces. On the other hand, the cube faces are small squares on the six solid angles of the octahedron. The angle between adjacent faces of a cube and an octahedron is 125° 16'.

Figure 18 shows a combination of the cube and dodecahedron, and Fig. 19 shows a combination of the octahedron and dodecahedron. Both the cube and the octahedron have twelve similar edges, and these are cut off equally, or truncated, by the twelve faces of the dodecahedron. In Fig. 20 is shown a crystal with the combination of the cube (*a*), octahedron (*o*), and dodecahedron (*d*). The angle between adjacent faces of the cube and the dodecahedron is 135°; between those of the octahedron and the dodecahedron it is 144° 44′.

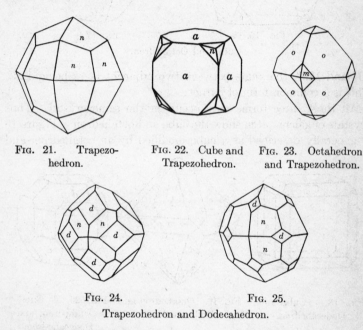

FIG. 21.    Trapezo-        FIG. 22.  Cube and      FIG. 23.  Octahedron
    hedron.              Trapezohedron.          and Trapezohedron.

FIG. 24.                          FIG. 25.

Trapezohedron and Dodecahedron.

**Trapezohedron.** A trapezohedron has twenty-four equal faces, each a four-sided figure or trapezium. Unlike the forms already described, which are always the same, there are several different trapezohedrons all having similar appearance but differing in the angles between the faces. It requires a much more extensive study than is possible or necessary for the beginner to learn how these forms are mathematically distinguished from each other. The trapezohedron in Fig. 21 is the most common and is frequently found on garnets.

Figures 22 to 25 show combinations of the trapezohedron (*n* or *m*) with the cube (*a*), octahedron (*o*), and dodecahedron (*d*). The last two are common combinations on garnet.

The trapezohedron is also called a tetragonal trisoctahedron because the form suggests an octahedron in which three faces take the place of a single octahedral face, each of them being a four-sided figure or tetragon.

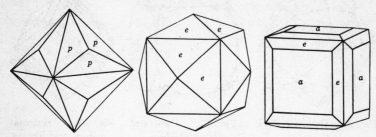

Fig. 26. Trisoctahedron.   Fig. 27. Tetrahexahedron.   Fig. 28.   Cube and tetrahexahedron.

There is also another trisoctahedron, called a trigonal trisoctahedron, shown in Fig. 26, which also has twenty-four faces, three of these also corresponding to an octahedron face; but each is a three-sided figure (trigon) or an isosceles triangle. This form does not often occur alone but may be seen on complex crystals of galena.

**Tetrahexahedron.** A tetrahexahedron (Fig. 27) has twenty-four faces, each face being an isosceles triangle and four together having the same position as the face of a cube. Figure 28 shows a combination of the cube and a tetrahexahedron; the latter is said to *bevel* the edges of the cube because the two planes are equally inclined to the two adjacent cube faces.

**Hexoctahedron.** A hexoctahedron (Fig. 29) is a forty-eight-faced solid; each face is a scalene triangle, and six faces have the same general position as a face of an octahedron.

Figure 30 shows a combination of the cube (*a*) with the hexoctahedron (*s*), as found on some fluorite crystals. Figure 31 is a common combination of forms in garnet with the hexoctahedron (*s*) beveling the edges of the dodecahedron (*d*).

Figure 32 (cuprite) and Fig. 33 (the rare species, microlite) show some rather complex combinations of the forms described. In Fig. 32 the cube (*a*) and the dodecahedron (*d*) predominate; the faces of the octahedron (*o*) are small; *n* and *β* are faces of two different trapezohedrons. In Fig. 33 the octahedron (*o*) predominates, the cube (*a*) is intermediate, and the dodecahedron (*d*) is subordinate; the faces *m* belong to a trapezohedron, and faces *p* belong to a trigonal trisoctahedron.

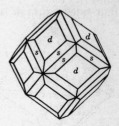

FIG. 29. Hexoctahedron.    FIG. 30.  Cube and    FIG. 31.   Hexocta-
                           Hexoctahedron.    hedron a1d Dodec-
                                                ahedron.

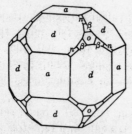
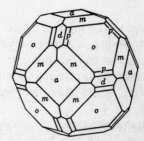

FIG. 32. Cuprite.           FIG. 33. Microlite.

**Tetrahedron.** The tetrahedron (Fig. 34) has four triangular faces, each of them an equilateral triangle. It may be considered as the half-form of the octahedron, since half the faces of the octahedron, if every other one is taken, will, if extended, form a tetrahedron (Fig. 35). Reducing the number of faces also reduces the symmetry, and in this symmetry class, the *hextetrahedral*, there are six planes, three 2-fold axes, four 3-fold axes, and no center.

The tetrahedron is the only common form in this class, but it is frequently found in combination with the cube as shown in Figs. 36 and 37. It is seen that the tetrahedron faces (*o*) are present only on the alternate angles of the cube. Figure 37 illustrates a combination of a cube and a tetrahedron in which the latter predominates.

Figure 38 shows a combination of the tetrahedron (*o*) with a similar form (lettered $o_1$) made up of the four remaining faces of the octahedron. It might be asked why this form cannot be regarded as an octahedron in which four faces are accidentally

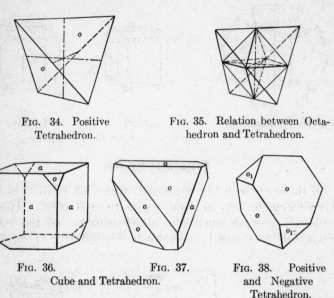

FIG. 34. Positive Tetrahedron.

FIG. 35. Relation between Octahedron and Tetrahedron.

FIG. 36.

FIG. 37.
Cube and Tetrahedron.

FIG. 38. Positive and Negative Tetrahedron.

larger than the others. This is impossible, for it can be shown, by differences of luster, that the eight faces are not *all* alike but make up two forms with four faces each, a combination of the positive and the negative tetrahedron. This, however, is a somewhat difficult subject for a beginner.

**Pyritohedron.** The pyritohedron (Fig. 39) is a twelve-sided solid, or dodecahedron, each face of which is a pentagon; it is thus sometimes called the pentagonal dodecahedron. In crystallography the name dodecahedron is usually given only to the rhombic dodecahedron which has been described (Fig. 15), and this form, the pyritohedron, takes its name from the mineral pyrite or iron pyrites, on which it is a common form.

The pyritohedron is the half-form of the tetrahexahedron and thus has a lower symmetry. It is in the *gyroidal* class with three 2-fold axes, four 3-fold axes, three planes, and a center of symmetry. The pyritohedron is the only common form in this class but frequently is found in combination with other forms. If in combination with the cube, the solids in Figs. 40 and 41 result; Fig. 42 is a combination of an octahedron and a pyritohedron.

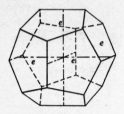

FIG. 39. Pyritohedron.          FIG. 40.                FIG. 41.
                                    Cube and Pyritohedron.

## Tetragonal System

All the crystals in the tetragonal system are referred to three crystallographic axes at right angles to each other. The two horizontal axes are equal and interchangeable, but the vertical axis is of different length. Because of this difference in length, the

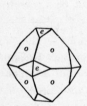  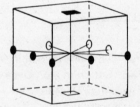  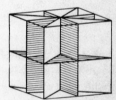

FIG. 42.    Octa-        FIG. 43.    Axes        FIG. 44.    Planes
hedron and Pyr-          Tetragonal Symmetry.
itohedron.

symmetry is lower than that of the isometric system. The symmetry class in which most tetragonal minerals fall is the *ditetragonal-dipyramidal* class having the symmetry: one 4-fold axis, four 2-fold axes, five planes, and a center, as shown in Figs. 43 and 44.

An examination of the drawings of tetragonal crystals will show that the grouping of the faces is the same about the ends of the horizontal axes but different from the grouping at the ends of the vertical axis. It will also be seen that about the vertical axis the faces of the same kind are arranged in fours, as in four pairs. In other words, the vertical axis is the one of 4-fold symmetry, and intersecting in this axis are four symmetry planes. Two of these planes include the horizontal crystallographic axes, and the other two planes lie between them at 45°. A fifth symmetry plane is at right angles to the others in the plane of the two horizontal crystallographic axes. All the four axes of 2-fold symmetry lie in the horizontal symmetry plane; two are the same as the crystallographic axes, and the other two are at 45°.

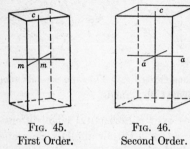
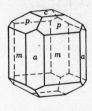

Fig. 45.          Fig. 46.          Fig. 47.
First Order.     Second Order.    First and Second Order.
                 Tetragonal Prisms.

**Tetragonal Prism and Basal Pinacoid.** The tetragonal or square prism (Fig. 45) has, like the cube, angles of 90° between the faces, but it differs from the cube in that there are four rather than six faces. The two end faces in Fig. 45 do not belong to the square prism but are *basal planes* belonging to the form known as the basal pinacoid and, therefore, differ from and are not interchangeable with the prism faces. This difference is often shown in the crystal by a difference in smoothness or luster of the two kinds of faces or by a cleavage parallel to one set of planes and not to the other.

In addition to the square prism just described, there is another one, similar but placed at 45° as shown in Fig. 46. Taken alone these two forms cannot be distinguished from each other, and it would be impossible for the beginner to tell which he was examining. However, if they occur together, the distinction is obvious. Thus Fig. 47 shows the two prisms on the same crystal, the faces of one truncating the edges of the other.

On a careful examination of Figs. 45 and 46 it will be seen that in the prism first mentioned (Fig. 45), called the *first-order prism*, the horizontal axes join the intersection edges of the prism faces.

In Fig. 46, the *second-order prism*, the horizontal axes join the midpoints of opposite faces. When only one of the prisms is present, it is customary to set it up as the first-order prism.

**Tetragonal Dipyramid.** The tetragonal dipyramid with its eight triangular faces looks somewhat like an octahedron, but here the faces are isoceles triangles rather than equilateral triangles, and

 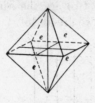 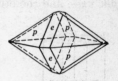

| FIG. 48. | FIG. 49. | FIG. 50. |
|---|---|---|
| First Order. | Second Order. | First and Second Order. |

Tetragonal Dipyramids.

the angle between two faces over a horizontal edge differs from that over one of the sloping edges. Either angle is characteristic of a given species and differs from one species to another. The crystal form known as a pyramid is found only at one end of the crystal. The term *dipyramid* as used here refers to two pyramids, one at each end. There may be several tetragonal dipyramids of the same type on one crystal but differing in their angles and consequently flatter or sharper at the apex.

Corresponding to the first-order tetragonal prism there is a first-order tetragonal dipyramid (Fig. 48), and, at 45° to this, there is a second-order tetragonal dipyramid (Fig. 49) corresponding to the second-order square prism. Here, again, it is impossible to distinguish between the two when only one type is present. However, when both dipyramids are found on the same crystal, the faces of one can be seen to truncate the edges of the other. Figure 50 shows the two dipyramids in combination.

It should be noted that the horizontal axes in the first-order prism meet the crystal at the point of intersection of four dipyramid faces; in the second-order dipyramid the same axes meet the midpoints of the intersection of a face below with a face above.

In Fig. 51 is shown the combination of a first-order prism with a second-order dipyramid; in Fig. 52 the reverse combination is shown. Figure 52 resembles a cube modified by an octahedron (compare with Fig. 16, p. 23), but it differs from it in that the angles between $p$ and $c$ are not equal to the angles between $p$ and $a$. Figures 53 to 55 represent combinations of forms found on crystals of zircon; Fig. 56 represents idocrase, and Fig. 57 rutile.

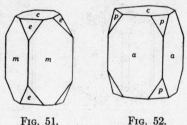

Fig. 51.           Fig. 52.

Combinations of Prisms and Dipyramids.

Besides the square prisms, there is the ditetragonal prism made up of eight similar faces. It is shown on the complex crystal represented in Fig. 60 with its faces lettered $h$. Corresponding to this prism there is the ditetragonal dipyramid, a double eight-sided

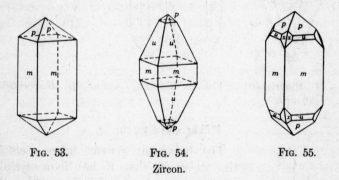

Fig. 53.        Fig. 54.        Fig. 55.

Zircon.

pyramid, as shown in Fig. 59. It is also shown in Fig. 60 lettered $z$ in combination with other forms. Figures 56 to 58 show some of the combinations of forms found on tetragonal crystals.

All the tetragonal forms that have been described belong to the symmetry given on p. 28. There are other forms of lower symmetry, but only one, the disphenoid, is common enough to be mentioned.

**Disphenoid.** The disphenoid (Fig. 61) is a four-faced solid resembling a tetrahedron but differs from it in that the faces are isosceles rather than equilateral triangles. It can be thought of as the half-form of the tetragonal dipyramid shown in Fig. 48. The mineral chalcopyrite commonly crystallizes in disphenoids that are very difficult to distinguish from tetrahedrons.

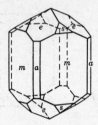
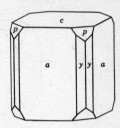

FIG. 56.  Idocrase.        FIG. 57.  Rutile.        FIG. 58.  Apophyllite.

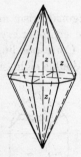
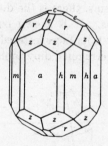
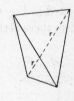

FIG. 59.  Ditetragonal        FIG. 60. Wernerite.        FIG. 61.  Disphenoid.
Dipyramid.

## Hexagonal System

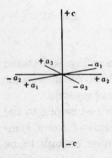

FIG. 62.    Hexagonal
Crystallographic
Axes.

The hexagonal system differs from all the others in that it has four crystallographic axes of reference (Fig. 62) rather than three. Three of these axes are of equal length and lie at 60° to each other in the horizontal plane. The fourth axis is vertical and is either shorter or longer than the horizontal axes. Although there are twelve symmetry classes in the hexagonal system, only two will be described here since most of the common hexagonal minerals are in these classes.

The class with the highest symmetry is the *dihexagonal dipyramidal*. It has seven planes, one 6-fold axis, six 2-fold axes, and a center of symmetry, as shown in Figs. 63 and 64. Three

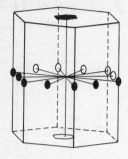

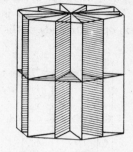

Fig. 63.  Axes.          Fig. 64.  Planes.

Hexagonal Symmetry.

of the 2-fold axes are coincident with the crystallographic axes, and the other 2-fold axes lie midway between them in the horizontal plane.  The 6-fold axis is coincident with the vertical crystallographic axis.  Six of the symmetry planes are vertical, and each includes one of the 2-fold axes;  the seventh plane is horizontal. On examination of the crystal drawings, one will see that this symmetry is shown by the 6-fold arrangement of faces about the end of the vertical axis and the 2-fold arrangement of faces about the ends of the horizontal symmetry axes.

**Hexagonal Prism and Basal Pinacoid.**  The hexagonal prism as shown in Fig. 65 is made up of six similar vertical faces, with angles of 120° between them.  This is an illustration of the *first-order prism* in which the crystallographic axes may be seen meeting the intersection edges of prism faces.  If we rotate the crystal 30° about the vertical axis, we will bring it into another position which satisfies the symmetry requirements equally well.  Here the crystallographic axes meet at right angles the center of each face, and in this position it is known as the *second-order prism* (Fig. 66).  These two prisms correspond to the two positions of the square prism in the tetragonal system.  If only one of these prisms is present on a given crystal, it is normally oriented as a first-order prism with a face toward the observer.  However, if both are present, as in Fig. 69, the best-developed one is set as the first-order and the other as the second-order.

It is obvious that, since the hexagonal prism is made up of only vertical faces, it cannot in itself form a solid. Another form must, therefore, always be in combination with it. In Figs. 65 and 66 the *basal pinacoid* is shown at the top and bottom of the crystals. The basal pinacoid is then made up of only two similar faces on opposite ends of the crystal.

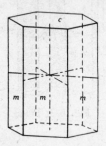
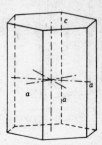

FIG. 65.   First Order.          FIG. 66.   Second Order.

Hexagonal Prisms.

**Hexagonal Dipyramid.** There are two hexagonal dipyramids corresponding to the two hexagonal prisms. These dipyramids are made up of twelve faces, each an isosceles triangle, six above and six below. The angles that the faces make with each other or with the prism faces are characteristic of a given species.

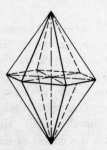
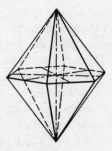
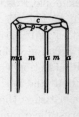

FIG. 67.   First Order.     FIG. 68.  Second Order.          FIG. 69.  Beryl.

Hexagonal Dipyramids.

Figure 67 illustrates the dipyramid of the first order, and Fig. 68 the dipyramid of the second order. If these figures are compared with Figs. 65 and 66, it will be seen that the faces of each of these pyramids lie directly above and below the faces of the corresponding prism. If a pyramid and a prism of different orders appear on the same crystal, it is customary to set the pyramid as first order. Figure 69, a beryl crystal, shows a combination of the first-order ($m$) and the second-order ($a$) prisms, first-order ($p$) and second-order ($s$) dipyramids, and the basal pinacoid.

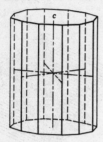
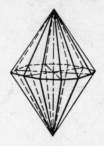

Fig. 70. Dihexagonal Prism.     Fig. 71. Dihexagonal Dipyramid

In addition to the prisms mentioned, there is a twelve-sided prism known as the dihexagonal prism. This form is made up of twelve similar vertical faces grouped in pairs at the ends of the three horizontal crystallographic axes (Fig. 70). Although the faces are similar in appearance, the alternate angles are different, and there are thus only the six vertical planes of symmetry.

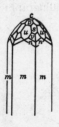
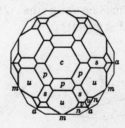

Fig. 72.     Fig. 73.

Beryl.

Corresponding to the dihexagonal prism there is a *dihexagonal dipyramid*, bounded above and below by twelve similar triangular faces. This form, shown in Fig. 71, is usually present only as small faces in combination with others. Figure 72 is a drawing of a beryl crystal in which the faces lettered $n$ and $v$ are dihexagonal dipyramids. Figure 73 is an enlarged map and shows the faces as they would appear on looking directly down on a beryl crystal much like that in Fig. 72.  Note also in Fig. 73 the first-order prism (*m*) and dipyramids (*u* and *p*), the second-order prism (*a*) and dipyramid (*s*), and the basal pinacoid (*c*).

There are several crystal classes in the hexagonal system having lower symmetry than the one just described, but it seems wise to mention here only one other, known as the *hexagonal-scalenohedral* class.

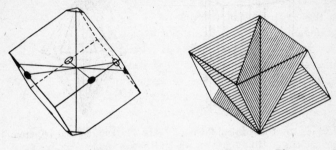

FIG. 74.  Axes.                    FIG. 75.  Planes.
Symmetry: Hexagonal-Scalenohedral Class.

The symmetry of the hexagonal-scalenohedral class is shown in Figs. 74 and 75.  There are, at 60° to each other, three vertical planes of symmetry, three 2-fold symmetry axes coinciding with the three horizontal crystallographic axes, one 3-fold symmetry axis coinciding with the vertical crystallographic axis, and a center of symmetry.

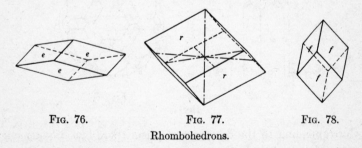

FIG. 76.                    FIG. 77.                    FIG. 78.
Rhombohedrons.

In the forms belonging to this class the faces are in threes or groups of three about the extremities of the vertical axis.

**Rhombohedron.**  The rhombohedron is a six-sided solid, each face of which is a rhomb, as shown in Figs. 76 to 78.  There are a great many rhombohedrons differing in the angles between the faces, and thus the general appearance may be flattened and obtuse (Fig. 76) or elongated and acute (Fig. 78). The rhombohedron looks somewhat like a cube with the line join-ing two opposite angles placed in a vertical position. In fact, the cube may be thought of as a special case of the rhombohedron with face angles of 90° and as coming between the obtuse and the acute rhombohedrons.

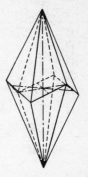

**Scalenohedron.** The scalenohedron (Fig. 79) is a twelve-sided solid, looking a little like a double six-sided pyramid, but the faces are sca-lene triangles and the edge is zigzag, up and down, like that of a rhombohedron, instead of horizontal as in the dipyramid. Moreover, only the alternate angles between the faces over the edges that meet in the vertex are alike;

Fig. 79. Scaleno-hedron.

in other words, there are two sets of three each, those of one set being more obtuse than those of the other.

The number of species crystallizing in the hexagonal scaleno-hedral class is very large, and some of them, as, for example, calcite, are highly complex. In the figures of calcite given here (Figs.

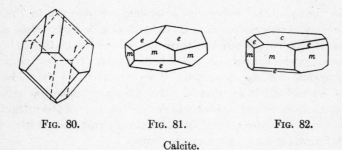

Fig. 80.  Fig. 81.  Fig. 82.

Calcite.

80 to 84) the faces $r$, $f$, $e$ belong to different rhombohedrons; $v$, to a scalenohedron; $m$ is the first-order hexagonal prism; $c$, the basal plane.

Figure 85 represents a more complex crystal, also of calcite, and Fig. 86 gives a basal projection of it. Here there are several rhom-bohedrons, $r$, $e$, $f$; the scalenohedrons, $y$ and $t$; the prism, $m$. These figures show well the symmetry about three planes meeting at angles of 60°. Figure 87 shows a crystal of hematite; $u$ and $r$

are faces of two rhombohedrons, and *n*, faces of a hexagonal dipyramid.

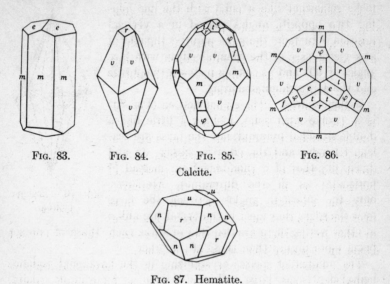

FIG. 83.      FIG. 84.      FIG. 85.      FIG. 86.

Calcite.

FIG. 87. Hematite.

## Orthorhombic System

The forms of the orthorhombic system are referred to three unequal crystallographic axes at right angles to each other, as shown in Fig. 88. There are three crystal classes in the orthorhombic system, but only the one of highest symmetry, the *rhombic-dipyramidal*, will be described. The symmetry of this class (Figs. 89 and 90) is as follows: three symmetry planes at right angles to each other, each of which includes two of the crystallographic axes; three axes of 2-fold symmetry coincident with the crystallographic axes; a symmetry center.

**Prism.** The orthorhombic prism is made up of four similar vertical faces, as shown in Fig. 91, in which the top face, lettered *c*, the basal pinacoid, is a rhomb, not a square, as in the square prism of the tetragonal system. The angle between two faces over one vertical edge is obtuse, or greater than 90°; the other angle is acute and just as much less than 90°. For instance, if the angles at the front and back edges are 100°, the angles at the two side edges would be 80°. There may be several prisms on crystals of the same species differing in the angles of the two

**edges,** but the angular difference between them is characteristic of the species.

**Dipyramid.** The orthorhombic dipyramid is shown in Fig. 92. It resembles somewhat the tetragonal dipyramid, but because of the difference in length of its horizontal axes its cross section is a

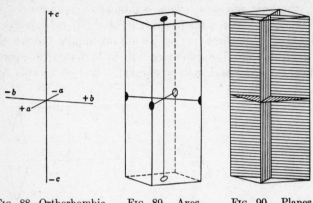

FIG. 88. Orthorhombic   FIG. 89. Axes.   FIG. 90. Planes.
Crystal Axes.            Orthorhombic Symmetry.

rhomb, not a square. Although the eight faces are similar, the edges between them belong in three sets, with different angles for each. There may be a variety of orthorhombic dipyramids, differing in their angles, and each corresponding to a given prism.

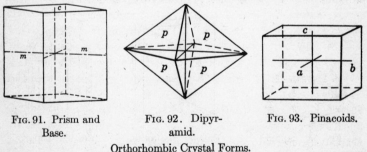

FIG. 91. Prism and   FIG. 92. Dipyr-   FIG. 93. Pinacoids.
Base.                amid.

Orthorhombic Crystal Forms.

In Fig. 99, *m* and *s* are prisms, and *e* and *f* are the corresponding dipyramids.

**Pinacoids.** The term *pinacoid* is familiar from the basal pinacoids of the tetragonal and hexagonal systems. They are forms made up of two faces, one at the top, the other at the bottom, of the crystal. The basal pinacoid has a like position in the orthorhombic system and is shown in combination with the prisms in Fig. 91. In Fig. 93 the basal pinacoid *c* is shown in combination with two other forms lettered *a* and *b*. These are also pinacoids since each form is made up of only two faces: *a*, the *front pinacoid*, and *b*, the *side pinacoid*. This combination resembles a cube since the angles between the faces are 90° but differs from it in that the faces belong in three sets, which are not similar to each other. Some crystals show the difference well in variation, in luster, or in etching, or a cleavage may be parallel to one set and not to the others.

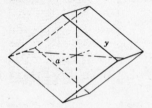

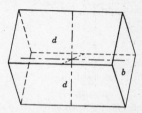

Fig. 94.   Brachydome and       Fig. 95.   Macrodome and
Front Pinacoid.                  Side Pinacoid.

**Domes.** The forms shown in Figs. 94 and 95 are called domes, from the Latin for house (*domus*) because, where they meet above, they make a horizontal edge like a roof. Domes, like prisms, are made up of four similar faces and, for this reason, are often called

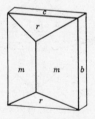

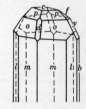

Fig. 96. Sulfur.          Fig. 97. Staurolite.          Fig. 98. Topaz.

*horizontal prisms.* There are two kinds of domes: the one parallel to the shorter horizontal axis, *a*, is the *brachydome* (Fig. 94); the other parallel to the longer horizontal axis, *b*, is the *macrodome* (Fig. 95).

Figures 96 to 98 show orthorhombic crystals with various combinations of forms. Figure 99 shows a complex crystal, and Fig.

100 is a projection of the same crystal on the base. Here the faces lettered *e, f* are two dipyramids; *d, h, k* are domes; *a, b* are front and side pinacoids, and *c* the basal pinacoid.

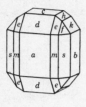

FIG. 99.

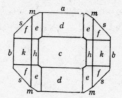

FIG. 100.

## Monoclinic System

In the monoclinic system there are three crystallographic axes of unequal length; two of them, the vertical and the right-left axis, are at right angles, but the third is inclined to the plane of the others, as shown in Fig. 101. The system is named *monoclinic* because of this one inclined axis.

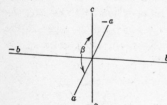

FIG. 101. Monoclinic Crystal Axes.

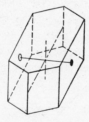

FIG. 102. Axis.

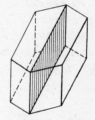

FIG. 103. Plane.

Monoclinic Symmetry.

The prismatic crystal class, the only one we shall consider in this system, has one plane of symmetry, one 2-fold axis, and a center (Figs. 102 and 103). The forms of the monoclinic system are similar to those in the orthorhombic, but the lower symmetry brought about by the inclination of an axis introduces some half-forms.

**Monoclinic Prism.** The monoclinic prism is made up of four similar vertical faces, as shown in Fig. 104, marked *m*. The angles formed at the intersections of the faces, as in the orthorhombic system, are not right angles, and several sets of prism faces may be present intersecting at different angles.

**Pinacoids.** There are three pinacoids in the monoclinic system, each of which is parallel to two of the crystallographic axes. In Fig. 105 the front pinacoid, *a*, and the side pinacoid, *b*, are at right angles and correspond to the front and side pinacoids of the orthorhombic system. Since the basal pinacoid, *c*, is parallel to the inclined axis, it slopes to the front with the *a* axis, and faces

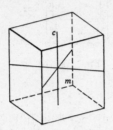

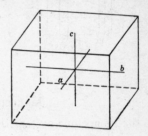

Fig. 104. Monoclinic Prism
and Base.

Fig. 105. Monoclinic
Pinacoids.

*a* and *c* are not at right angles. The slope of the basal pinacoid is always the same for a given species but differs from one mineral to another. The angle made by the intersection at *a* and *c* at the front of the crystal is obtuse, but the corresponding angle at the back of the crystal is acute. The difference in these angles gives rise to the half-forms mentioned below.

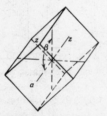

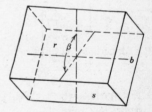

Fig. 106. Clinodome.

Fig. 107. Orthodomes.

**Domes.** We saw in the orthorhombic system that a dome is parallel to one of the crystallographic axes and intersects the other two. In the monoclinic system the form parallel to the inclined axis is the *clinodome* and is composed of four similar faces (Fig. 106). The dome parallel to the right-left axis is the *orthodome*. Since the faces of this form must lie over the alternately obtuse and acute edges between *a* and *c* (Fig. 105), all of them cannot be similar. There are thus two orthodomes, each a half-form composed of two faces. The positive orthodome (*r*, Fig. 107) lies over the obtuse edges, and the negative orthodome (*s*, Fig. 107) lies over the acute edges. One of these half-forms may be present without the other.

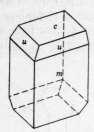
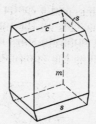

FIG. 108. Positive.    FIG. 109. Negative.

Monoclinic Pyramids.

**Pyramids.** The pyramids in the monoclinic system are, like the orthodomes, half-forms, since all the corners or edges over which they lie are not similar. If we consider Fig. 104, we can see that the angles at the top front and bottom rear are obtuse, whereas the remaining angles are acute. There must therefore be two pyramids: the *positive pyramid* (*u*, Fig. 108) lies over the obtuse

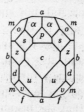

FIG. 110.    FIG. 111.    FIG. 112.    FIG. 113.

Monoclinic Crystals.

edges, and the *negative pyramid* (*s*, Fig. 109) lies over the acute edges. These two forms, each with four faces, are quite independent, and one of them is commonly found without the other.

Figures 110 to 113 are drawings of various monoclinic crystals with the forms lettered as follows: *m* and *f*, prisms; *u* and *v*, positive pyramids; *o* and *s*, negative pyramids; *a*, *b*, and *c*, pinacoids. Figure 113 shows a basal projection of a more complex crystal; here the symmetry parallel to the *b* faces is clearly exhibited.

## Triclinic System

In the triclinic system the faces are referred to three unequal axes, all oblique to each other. Planes and axes of symmetry are lacking, and a center is the only element of symmetry present. Like pinacoids all the forms of the triclinic system have only two faces on opposite sides of the crystal. For this reason the class is known as the *pinacoidal crystal class.*

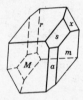          

Fig. 114.  Axinite.          Fig. 115.  Albite.

Triclinic Crystals.

Figure 114, axinite, and Fig. 115, albite feldspar, show two triclinic crystals. Here it is seen that the like planes are in sets of two each—one in front, the other behind, represented in dotted lines. In Fig. 115 there is some resemblance to a monoclinic crystal, but the angle between the faces *b* and *c* is not 90°, as it must be there; and, moreover, the angles *bm*, *bM* are different, as are also the angles *bo*, *bp*. Hence, *m* and *M* are different planes, and also *o* and *p*. The subject of triclinic crystals will not be carried further, because of its great complexity. Fortunately only a few common minerals crystallize in the triclinic system, so that the beginner is not often confronted with study of them and the problem of their orientation.

## Measurement of Angles

It is helpful in studying minerals to be able to measure the angles between the faces of a crystal, for in this way it is possible to tell a square prism (tetragonal) from a rhombic prism (ortho-rhombic), a cube from a rhombohedron, or an octahedron from a dipyramid. The simplest method, if the crystal is large enough, is to use a contact goniometer, as shown in Fig. 116. The crystal

is placed between the jaws as shown, so that the two faces whose
angle is to be measured are in contact with the jaws, and the edge
between these faces is at right angles to them. The angle is then
read from the scale.

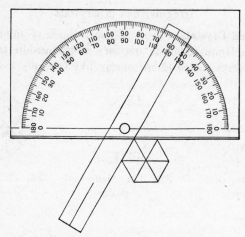

Fig. 116. Contact Goniometer.

A fair substitute for an expensive goniometer can be made
from a cheap protractor; two arms of thin wood or plastic, shaped
like the steel ones of Fig. 117, are cut out and then a brass pivot
on which the arms can turn is put
through them. It is not necessary
or desirable that the arms be permanently attached to the protractor.
One pair of edges (the inner edges to
the right in the figure) must be exactly in line with the center or pivot;
between them the angle is read off
when the arms are placed on the pro-

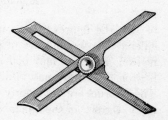

Fig. 117. Arms for Contact
Goniometer.

tractor, the pin then passing through
its center and one edge through its
zero. The other pair of edges (to the left) must be parallel
to those mentioned first, so that they give the same angle; the two
faces of the crystal whose angle is to be measured are placed between them, as already explained.

For the accurate measuring of angles or for those of a very small crystal a reflecting goniometer is required. This instrument, described in larger works, is expensive, and its use requires both skill and experience. It demands, moreover, that the crystal faces be of high quality.

## Irregularities of Crystals

**Malformed Crystals.** Most of the crystals of minerals would give a poor impression of nature's workmanship to one who expected always to see them exactly like carefully made models,

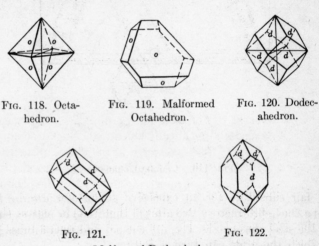

FIG. 118. Octa-         FIG. 119. Malformed         FIG. 120. Dodec-
hedron.                   Octahedron.                 ahedron.

FIG. 121.                                  FIG. 122.

Malformed Dodecahedrons.

or like the figures given on the preceding pages. The cubes of galena that we find are often flattened or drawn out. An octahedron (Fig. 118) may be flattened to look like Fig. 119; a dodecahedron (Fig. 120) may take the forms shown in Figs. 121 and 122. These forms are not poorly made, like a poor model; on the contrary, the size of the like faces on a crystal may vary, and therefore the shape of the solid as a whole may vary, but *the angles* between them remain the same. Moreover, when we study a crystal more carefully, we find that what is really essential is, not the size or shape of each face, but the way in which the atoms are arranged. For example, in a cube the fact that the structure is the same in the three directions at right angles to the cube faces is an essential point. It follows that in the cube not only are the angles between two adjacent faces always 90°, but also the six cube faces are all similar. Therefore, if there is easy fracture, called cleavage, parallel to one cube face, there will be the same cleavage parallel to the others. Moreover, the etching and luster on all the cube faces will be the same, and the actual size of the faces is a matter of no importance. In fact, in one species the cubes are sometimes lengthened so that they are like fine hairs.

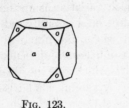

FIG. 123.          FIG. 124

Cube and Octahedron.

Similar remarks can be made in regard to the malformed octahedron and dodecahedron already illustrated, and indeed about any malformed crystal. The symmetry in the atomic structure, and hence the angles between the faces, remain unchanged, although the external geometrical symmetry is not that of the ideal model.

Another good example of what is possible in a malformed crystal can be explained by referring to Fig. 123, a cube with octahedral faces on its solid angles. Instead of this ideal form, it is common to find in natural crystals no two of the triangular octahedral faces of the same size; some of them may even be absent; the cubic faces also vary, Fig. 124. But such a crystal is not essentially different from Fig. 123, for every octahedral face is identical with each of the others if it is equally inclined to the three adjacent cube faces, that is, to the three crystallographic axes, even if all the faces differ in size. In other words, it is always the position of the face, not its size, which is essential.

In the same way a cube may in nature look like a square tetragonal prism, for all the angles between the faces are right angles in both cases, and the goniometer will not tell the difference between them, as has been already explained; but the atomic structure of the two is not to be confused. In the tetragonal prism there is the same arrangement in the horizontal directions, but a different arrangement along the vertical direction. Hence the square top of the crystal appears different from the four vertical faces, and we may find cleavage parallel to one set and not to the other. For example, the mineral, apophyllite forms pseudo-cubes in combination with a dipyramid which resembles a cube and octahedron (Fig. 125), but there is a cleavage parallel to the base only. Moreover, there is a pearly luster on the base and a high glossy luster on the faces of the square prism.

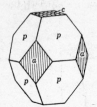

FIG. 125. Apophyllite.

Because so much variation is possible in the size of the like faces on a crystal, and hence in the shape of the whole, the practical study of natural crystals is much more difficult than the study of the models which give the ideal geometrical symmetry. A careful comparison of the well-developed crystals in Fig. 126 with the wooden models that represent them will show that it is the exceptional crystal that is a perfect geometrical solid. Most crystals are so implanted on the rocks or embedded in them that only part

FIG. 126A.   Crystals.

*A.* Calcite, *B.* Quartz, *C.* Barite, *D.* Fluorite, *E.* Galena, *F.* Tourmaline, *G.* Pyrite, *H.* Garnet, *I.* Orthoclase, *J.* Beryl, *K.* Idocrase.

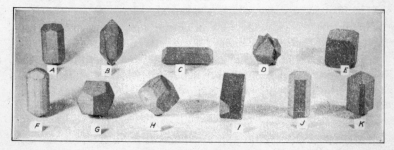

FIG. 126B.   Models.

of the crystal has been developed. Thus a quartz crystal is often attached at one extremity, and only the other end has had a chance to grow freely. Or crystals may be implanted upon a surface of rock so that only a series of minute faces and angles is visible. In that event the study of the crystal is really a difficult matter requiring much skill and experience, and the beginner should not be discouraged because he cannot at once tell the form of a crystal. Even here, however, some conclusion can often be drawn from the shape of the faces; thus, if they are equilateral triangles, they probably belong to an octahedron; if rhombs, to a rhombic dodecahedron, and so on.

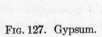

FIG. 127. Gypsum.    FIG. 128. Beryl Crystal broken and cemented by Quartz.

Besides the malformed crystals just considered, which, although they look irregular, are really perfect in regard to the position of the faces and the angles between them, there are others which are really deformed. Some unusual conditions attending the growth of the crystal or perhaps some force which has acted upon it since it was formed may have bent or twisted it out of its normal shape. Such crystals vary not only in the size and the shape of the faces but also in the angles between the faces. Thus the faces may be curved, as in the barrel-shaped crystals of pyromorphite, or convex as is common in crystals of the diamond; or the whole crystal may be bent, like some crystals of quartz or stibnite, or some kinds of chlorite and gypsum. (See Fig. 127.)

Aside from this curving and twisting, a crystal may have had its shape more or less changed by a force exerted in the rock since it was made. It may even have been broken and later cemented together, so that many irregularities may result. Figure 128 shows a crystal of beryl which has been broken into many pieces; these have been slightly displaced from each other, and the whole has been cemented together by quartz.

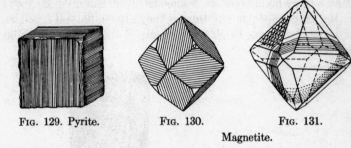

FIG. 129. Pyrite.    FIG. 130.    FIG. 131.

Magnetite.

Other irregularities of crystals besides those mentioned occur. The faces of crystals, instead of being perfectly smooth, are often rough, perhaps because they are made up of a multitude of points. Or they may be covered with fine lines, or *striations*, like those on the cubic faces of pyrite (Fig. 129), which are explained by the successive combination of another face (the pyritohedron) in narrow lines with the cube face. In the magnetite crystal (Fig. 130) the fine lines represent striations on a dodecahedral face due to the presence of the octahedron also. This *oscillatory combination*, as it is called, may even make the crystal nearly round, like some prismatic crystals of tourmaline. Striations may be due to twinning, as is common with the triclinic feldspars. (See p. 55; also Fig. 329, p. 237.) Figure 131 shows an octahedral crystal of magnetite with twinning lamellae appearing as striations on an octahedral face.

Other crystals may have faces with a multitude of little elevations or depressions, the latter like the pits spoken of on p. 17 as having been produced by etching; in fact they can sometimes be explained as etching by nature. The same cause—the action of some partial solvent after the formation of the crystal—often explains the rough faces to which we have alluded. The careful examination of some crystals may show the replacement of a face by two or more others varying a little from it in angular position.

The four slightly raised faces that take the place of the cube face on some English fluorite crystals are good examples. Such planes are often called *vicinal* planes (from the Latin *vicinus*, neighboring).

Crystals which have formed rapidly may have only a more or less regular skeleton shape, like the crystal of salt represented in Fig. 132, and the drawing on p. 182. Some salt crystals show one face distinctly but with a depression in the center, so that they are called hopper-shaped crystals. The cavernous crystals of pyromorphite and vanadinite are other examples. Crystals often enclose foreign substances of both solid and liquid material. Some quartz crystals contain water, occasionally with a movable bubble of gas. In others the

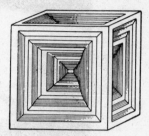

Fig. 132. Halite.

liquid may be carbon dioxide, with a bubble of the same substance in the form of gas. Such a crystal must have been formed under great pressure, sufficient to keep the gas in the liquid form. Fragments of such crystals heated in the gas flame fly to pieces with great violence, because of the expansion of the gas formed from the liquid by heat.

More commonly, crystals contain foreign solid substances of many kinds; quartz crystals enclose clay, particles of carbon, chlorite, rutile, tourmaline, and others. The famous groups of calcite crystals from Fontainebleau and from the Bad Lands of South Dakota (Fig. 133) contain some 60 per cent of quartz sand. It is most remarkable that the force of crystallization was powerful enough under such circumstances to marshal into place the atoms forming the calcite. In some crystals impurities are regularly arranged, and a curious effect is obtained in a cross section if it is cut and polished. Figure 134, garnet enclosing quartz, shows such an effect. The variety of andalusite called chiastolite affords another interesting example. Sections obtained from different crystals appear in Fig. 135.

**Pseudomorphs.** The word pseudomorph means *false form*, and the name is applied to a specimen having the form characteristic of one species and the chemical composition of another. This seeming contradiction is easily explained. Most chemical compounds are liable to undergo a change or alteration when subjected to certain conditions, such as moisture, the action of alkaline waters, or acid vapors. Thus, the mineral cuprite ($Cu_2O$) is rather easily changed chemically to malachite, the carbonate of

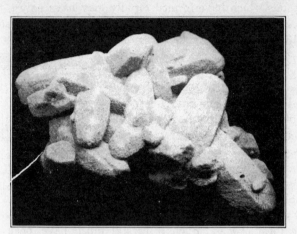

FIG. 133.   Sand Crystals, South Dakota.

copper.   Anhydrite ($CaSO_4$) assumes water and changes to the hydrated sulfate, gypsum ($CaSO_4 \cdot 2H_2O$); pyrite ($FeS_2$) changes to the hydrated oxide, limonite, $FeO(OH) \cdot nH_2O$.

FIG. 134.   Garnet enclosing
Quartz.

FIG. 135.   Chiastolite.

Now, in these and similar minerals, if the original specimen was in crystals the external form may be perfectly preserved, although the chemical composition and the atomic structure have changed. Hence, we describe the false forms mentioned as pseudomorphs of *malachite* after *cuprite; gypsum* after *anhydrite; limonite* after *pyrite.*

Other examples are pseudomorphs of chlorite after garnet, pyromorphite after galena, kaolin after feldspar. In a few rare instances, where the same chemical compound occurs in nature in two distinct crystalline forms, each with its own atomic structure, a change may take place in the structure without alteration of the chemical substance. Thus the rare mineral brookite ($TiO_2$) may be changed to rutile (also $TiO_2$). Such pseudomorphs have the special name of *paramorphs.*

The cases in which the original substance has entirely disappeared and some other has come to take its place are also called pseudomorphs. Thus we occasionally find quartz in the form of calcite, or of fluorite, or of barite, that is, a pseudomorph after one of these; also cassiterite in the form of orthoclase feldspar; native copper in the form of aragonite. Even fossil wood may be said to be a pseudomorph of quartz or opal after the original wood, the structure of which it sometimes preserves with wonderful perfection.

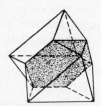

Fig. 136. Twinned Octahedron (spinel twin).

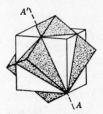

Fig. 137. Fluorite Penetration Twin.

## Groupings or Aggregations of Crystals

When crystals occur alone the forms are usually developed on all sides with some of the regularity of the ideal model. Thus perfect garnets are found in mica schist or granite, and gypsum crystals are found in clay. But it is still more common to find crystals grouped together either irregularly, as in the majority of cases, or perhaps in parallel position, or again in the peculiar way called twinning; the last will be described first.

**Twin Crystals.** The most interesting and important type of crystal grouping is shown in those complex forms called *twins.*

Figure 136 represents a twinned octahedron, and Fig. 137 represents two twinned cubes. In the first example the growth of the crystal as a whole has been such that one half has been developed in reversed position to the other, as if it had been revolved through 180° about an axis (called the twin axis) at right angles to two opposite octahedral faces. This is described as a *contact twin*.

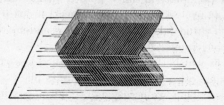

FIG. 138. Twinning by Reflection.

In Fig. 137 the two cubes interpenetrate each other, but each one is in such a position that it too appears to have been revolved

FIG. 139. Cas-
siterite.

180° around an axis running through two diagonal angles (the same octahedral axis as in the first example). This is called a *penetration twin*.

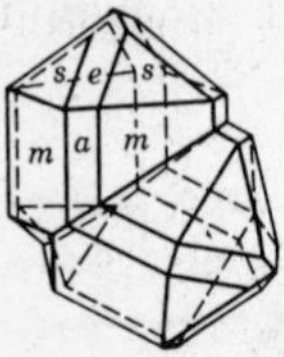

These two examples illustrate the most essential features of a twin. In each the two crystals, or parts of crystals, are in such a position that one seems to have been turned 180° with reference to the other and usually about an axis at right angles to some simple crystal plane, which is called the *twin plane*. Contact twins can be thought of as though one part were derived

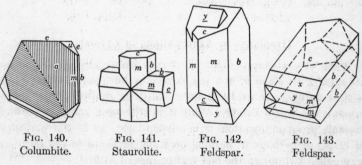

FIG. 140.
Columbite.

FIG. 141.
Staurolite.

FIG. 142.
Feldspar.

FIG. 143.
Feldspar.

Examples of Twinning.

from the other by reflection across a plane, as in the illustration of the gypsum twin (Fig. 138).

Figure 139 shows a contact-twin crystal of cassiterite or tinstone; Fig. 140, one of columbite, which also illustrates the point that the difference in direction of the striations of the two halves shows that the crystal is twinned. Figure 141 is a penetration twin of staurolite. Figures 142 and 143 show twins in feldspar. Figures 144, 145, and 146 show what are called repeated twins, which are often very regular. The twin aggregate may be made up of perhaps three, five, six, or even eight parts of crystals or complete crystals symmetrically arranged to resemble a star in many instances. Figure 147, rutile, shows a different kind of repeated twinning.

| FIG. 144. | FIG. 145. | FIG. 146. | FIG. 147. Rutile |
|---|---|---|---|
| | Repeated Twins. | | Twin. |

Another type of twinning yields thin parallel layers, each in reversed position to the next. It is called *polysynthetic* twinning and is best illustrated by a piece of a triclinic feldspar showing fine lines or striations on a surface of basal cleavage; these lines are simply the edges of the thin successive parallel plates. If the specimen is held so as to catch the reflection of light from a distant window and is turned through a very small angle, first one set of edges reflects and then the other set. This type of twinning is illustrated by a figure of albite on p. 237. Figure 131 on p. 50 shows polysynthetic twinning lamellae in a crystal of magnetite.

It must be understood that in most specimens showing twinning the revolution spoken of has not actually taken place. The rule is simply given in this form to show best the mathematical relations in position of the two parts. Still it is most interesting to note that in a few instances it is possible to cause the atoms of part of a crystal to change their position and produce twinning artificially, as, for example, by pressure in the proper direction.

Thus Fig. 148 represents a cleavage piece of calcite placed with an obtuse edge on a firm surface and then pressed by a knife (not too sharp) at *a*. Steady uniform pressure serves to reverse the position of the atoms in the part lying to the right so that the whole is pushed to the side and assumes a twinned position with reference to the rest. If the pressure is skillfully applied, no change in the transparency of this part takes place, and the new surface *gce* is perfectly smooth. It should be noted that to perform this experiment successfully clear calcite known as *Iceland spar* should be used. In nature, pressure may have pro-

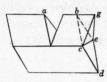

FIG. 148. Calcite. Artificial Twinning.

duced twinning after the formation of the crystal; it is then called *secondary twinning*. The twinning layers or lamellae observed in most cleavage masses of otherwise clear calcite may often be explained in this way. The similar layers which are common on large crystals of pyroxene and which cause a separation or "parting" parallel to the basal plane, appearing much like the easy fracture called cleavage, are caused by twinning.

It is evident that, since the crystals on a single specimen of a species may be grouped in a great variety of ways, it is not always easy to decide whether a given intergrowth is a twin or not; the decision often requires careful study, exact measurement of angles, and optical examination. For example, it is common to find quartz crystals crossing each other at a great variety of angles, but real twins like that in Fig. 149, known as a *Japanese twin*, are rare.

The twin plane cannot be a crystal symmetry plane nor the twin axis a 2-, 4-, or 6-fold symmetry axis. If a crystal were reflected across a symmetry plane, there would be no change in the orientation of the two halves. The twin plane is, therefore, parallel to some common crystal face that is not a plane of symmetry. Similarly, turning a crystal 180° about an even-fold symmetry axis will return it to the original position.

**Parallel Grouping.** One common type of crystal grouping, one which the beginner is likely to confuse with twinning, is parallel growth. In such aggregates crystals or parts of crystals are parallel to each other, so that the axes of all have the same directions and are not inclined as in most twins. This is illustrated by a pile of cubes with faces parallel and having re-entrant angles between them. Some crystals of many species are arranged in this way, but in every case it will be found that, if the group is held so that it reflects the light, the faces on adjoining crystals which reflect at the same time are always similar faces. An octahedron of fluorite, built up of a multitude of little cubes in parallel position, is a common example. Figure 150 shows a complex crystal of analcime, formed of a number of single crystals all of which are parallel; it is hence *not* a twin. Figure 151, quartz, illustrates well this parallel grouping in the subordinate parts.

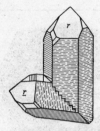

FIG. 149. Quartz, Japanese Twin.

FIG. 150. Analcime.

Parallel grouping is most interesting when the result is an aggregate built up with branching and rebranching parts like the

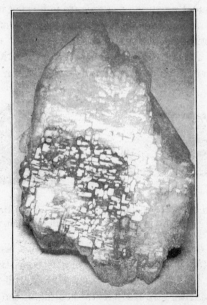

FIG. 151. Quartz. Parallel Growth. Greenwood, Maine.

limbs of a shrub or tree, and forming what are called arborescent or dendritic groups. Here all the crystals or parts of crystals

have like axes in the same direction.   This is shown in Fig. 152,
a crystal of native copper.   In Fig. 153 the little plates of hematite
are grouped together with such variation in their positions that the
top of the aggregate has the shape of a rosette.   Such a crystal is
called by the Germans *Eisenrose*, or iron rose.

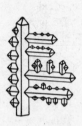

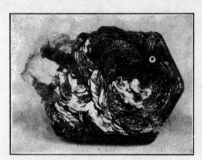

FIG. 152.   Native        FIG. 153.   Hematite, *Eisenrose*.   St.
Copper.                          Gothard, Switzerland.

Another interesting type of aggregate is formed when a number
of crystals are implanted upon the surface of another which has
obviously so influenced their growth that they are in parallel
groups and in a definite position relative to it.   Many examples
have been noted, such as chalcopyrite crystals on sphalerite and

        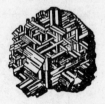

FIG. 154.   Rutile on        FIG. 155.   Rutile
Hematite.                          after Hematite.

rutile crystals on a tabular crystal of hematite, as shown in Fig.
154.   Figure 155 is a related example; it consists now of rutile
alone and has been described as a pseudomorph (see p. 52) of
rutile after hematite.   In the natural healing of the broken surface
of a crystal, such as quartz (alluded to on p. 19), it follows, almost as a matter of course, that the new growths are oriented in
directions parallel to the old ones.

**Mineral Aggregates.** Besides the twins and the parallel growths already mentioned, mineral aggregates are common in which the individuals that make them up are not crystals as we have learned to regard them. In fact, in most mineral collections at least half of the specimens will show no crystal faces and are said to be *massive*. There are, however, important distinctions of habit among massive minerals.

It has been pointed out that the presence of cleavage in a mineral is proof of its crystallinity. The cleavage may be in the same direction in all parts of the specimen as if it were a fragment of one large crystal, but more commonly cleavage direction changes, showing that the specimen is made up of many grains with different orientations. In such a specimen the mass is obviously crystalline, and the aggregate is said to be *cleavable* and *granular*. Such a mass of galena is really made up of a multitude of little grains, each one of which has its own directions of cleavage and presents to the eye its own edges. If the individual grains are large, the aggregate is said to be *coarse-granular;* if small, it is *fine-granular*. From the latter we pass to the closely *compact* kinds in which the state of aggregation may not be at all obvious to the eye; they may then be said to be *impalpable*. But this extreme is rare; for example, a piece of white marble, even if it is so fine-grained that the particles cannot be seen by the eye, usually sparkles in a strong light from the reflection of the multitude of minute cleavage faces.

This granular texture may belong also to other minerals which do not have cleavage as evidence of crystalline structure, and then the grain boundaries are difficult to see without the aid of polarized light. In some granular kinds of pyroxene, the small grains are found to be imperfect crystals.

*Lamellar* is a term applied to a mass made up of layers, whether separable or not. If the mass is in thin leaves or plates which can be separated from each other, it is called *foliated*, as with graphite. It is further called *micaceous* when the separation takes place as readily as in a piece of mica.

If the mass is made up of little columns, it is called *columnar*, as in some specimens of calcite. When there are distinct fibers, the mass is said to be *fibrous*, as in asbestos, in which the fibers are easy to pull apart, or *separable*. There are many intermediate kinds between *fine-fibrous* and *coarse-columnar*.

If all the fibers, or little columns, or leaves go out from a center like the spokes of a wheel, the aggregate is said to be *radiated* (cf. wavellite, Fig. 156) or perhaps *stellate* when it is star-shaped. If

FIG. 156.   Wavellite, Hot Springs, Arkansas.

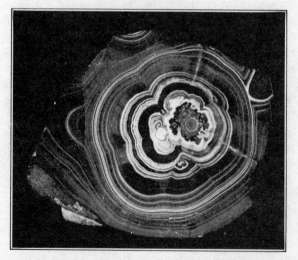

FIG. 157.   Malachite, Ural Mountains, Russia.

the layers are arranged in parallel position about one or more centers, the aggregate is said to be *concentric* (cf. malachite, Fig. 157). All these terms, granular, foliated, lamellar, columnar, fibrous, radiated, and concentric, ordinarily refer to the aggregate when it is more or less distinctly crystalline.

When the external surface of the mineral is in the form of rather large rounded prominences, it is called *mammillary* (cf. hematite, Fig. 158); if the prominences are smaller, somewhat resembling a bunch of grapes, it is called *botryoidal* (cf. prehnite, smithsonite,

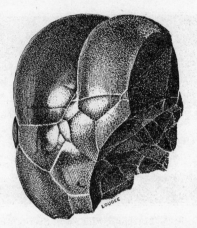

FIG. 158. Mammillary Hematite.

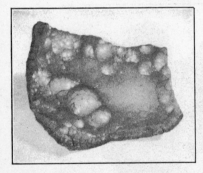

FIG. 159. Botryoidal Chalcedony, Desolation Island, Southern Chile.

FIG. 160. Reniform Hematite, England.

and chalcedony, Fig. 159). If the surface is made up of little spheres or globules, it is called *globular* (cf. hyalite or prehnite). If the surface resembles that of a kidney, it is called *reniform* (cf. hematite, Fig. 160). It should be understood that there is no sharp line dividing these different surfaces, and thus the term *colloform* is used to include all of them.

Some minerals take the form of a delicate, branching coral, and

they are called *coralloidal*, such as certain varieties of aragonite
known as *flos ferri* (Fig. 161).  If they are made up of forms like
small stalactites, they are said to be *stalactitic* (as limonite, Fig.
162).

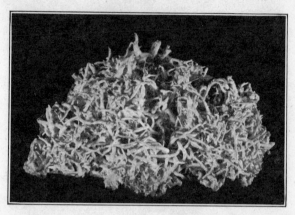

FIG. 161.   Aragonite, *Flos Ferri*, Styria, Austria.

FIG. 162.   Stalactitic Limonite, Ural Mountains, Russia.

If the material has clustered about a center like those curious
forms of impure calcium carbonate, called concretions, common in
clay (Fig. 163), it is said to be *concretionary*.  Dendrites, or
dendritic forms, are those which have more or less the shape of a
branching shrub or tree, like the forms of manganese oxide (see

Fig. 164) often seen on surfaces of smooth limestone or enclosed in moss agates. Some dendritic forms are made up of little crystals grouped together in parallel position. (See p. 57.)

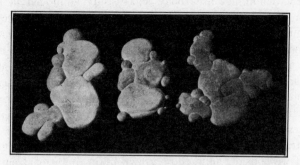

Fig. 163. Concretions.

Fig. 164. Dendrites of Manganese Oxide.

## Crystal Growing

For the beginner in the study of mineralogy, much of the discussion of crystallography on the preceding pages may seem very abstract and not pertinent to the few mineral specimens he has in his collection. Working with wooden or paper models may help, but, to understand crystals, one should work with and handle crystals themselves. Anyone can get good real crystals that show many of the forms described by growing them himself! These will not be minerals, as is pointed out on p. 4, but they will be just as good for studying crystallography.

Many techniques are today successfully employed in crystal growing, but the easiest for the beginner is the growing of crystals of soluble salts from water solutions of these salts. Such experiments are not difficult to carry out, and, if directions are followed carefully, one will be rewarded with well-formed crystals large enough to handle and study. Moreover, in watching crystals develop from tiny nuclei, one gets some insight into the principles of crystal formation as well as an appreciation of the great variety of characteristics that crystals of different substances have.

A large number of factors influences the generation of crystals even if we consider only those grown by evaporation from water solutions. Some of these factors are characteristic of the salt and are not controllable, but others can be controlled by the experimenter. The more important controllable factors are:

1. Temperature of the solution.
2. Variations in the temperature of the solution.
3. Rate of evaporation of the solution.
4. Dust particles in the solution.
5. Impurities dissolved in the solution.
6. Shape of the dish.

Other factors not controllable are:

7. Solubility of the salt.
8. Degree of supersaturation tendency of the salt.
9. Ability of the salt to crystallize.

If we are to grow large crystals of salts, we must try to adjust our conditions so that all these factors are favorable. Numerous as these factors are, there are many substances for which such control is easily possible. Among them are the following: copper sulfate, $CuSO_4 \cdot 5H_2O$; potassium alum, $KAl(SO_4)_2 \cdot 12H_2O$; potassium ferrocyanide, $K_4Fe(CN)_6 \cdot 3H_2O$; potassium ferricyanide, $K_3Fe(CN)_6$; barium chloride, $BaCl_2 \cdot 2H_2O$. Figures 165 to 169 illustrate well-formed crystals of the above compounds. Copper sulfate is triclinic and of a deep azure-blue color; potassium alum is isometric and forms in colorless octahedrons; both potassium ferrocyanide and potassium ferricyanide are monoclinic; the former is lemon yellow and the latter ruby red; barium chloride forms in colorless monoclinic crystals. These salts are soluble, but not excessively so; they do not supersaturate their solutions easily, and they crystallize readily, forming large perfect crystals without difficulty. The procedure for growing these crystals will be directed toward controlling the first six factors listed above.

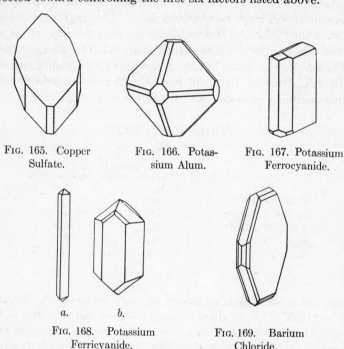

FIG. 165. Copper Sulfate.

FIG. 166. Potassium Alum.

FIG. 167. Potassium Ferrocyanide.

*a.* *b.*

FIG. 168. Potassium Ferricyanide.

FIG. 169. Barium Chloride.

Artificial Crystals.

Two techniques can be used. In one, the crystals may be grown on the bottom of a flat-bottom dish. This arrangement depends on a convection flow about the crystal, thus: 1. The solution at the air surface loses water by evaporation and becomes saturated or slightly supersaturated. 2. It thus becomes denser and sinks to the bottom of the dish. 3. Here it loses some salt to the growing crystal. 4. Becoming lighter, the solution rises to the top again. The second technique consists of suspending the crystal on a wire in the center of the solution. Under these conditions the cycle described above does not operate, and the solution must be constantly and gently stirred. Although this technique gives somewhat better crystals, the first is less trouble to arrange and gives very satisfactory results.

The general method for growing large crystals should follow these steps and considerations:

1. *Preparation of the solution.* Before a solution can deposit material on a crystal, it must be saturated, that is, the maximum amount of salt must be dissolved in it. This maximum amount (called the "solubility") is a definite quantity for any salt at any given temperature. These quantities are listed for a large number of compounds in many standard chemical tables and handbooks. The solubilities of the salts mentioned are given in the table. Since solubility increases with temperature, the easiest way to be

TABLE OF SOLUBILITY

| | SOLUBILITY IN WATER (grams per liter) | |
|---|---|---|
| SALT | 20° C (cold) | 100° C (boiling) |
| $CuSO_4 \cdot 5H_2O$ | 318 | 1158 |
| $KAl(SO_4)_2 \cdot 12H_2O$ | 108 | 1000 at 80° |
| $K_4Fe(CN)_6 \cdot 3H_2O$ | 322 | 940 |
| $K_3Fe(CN)_6$ | 442 | 775 |
| $BaCl_2 \cdot 2H_2O$ | 419 | 687 |

sure that a solution is saturated at room temperature (between 20° and 25° C) is to dissolve a quantity of salt in a liter of hot water about 25 per cent in excess of the solubility at room temperature and allow the hot solution to cool overnight. The next day the excess salt will have "showered" out of the solution in small crystals, and the clear solution above the deposit will be truly saturated. This will be the "stock solution," and it should always be kept in the presence of excess salt.

2. *Preparation of the crystallizing bath.* The dish should be broad and flat-bottomed. "Crystallizing dishes" of various sizes (4 inches in diameter is convenient), available at laboratory supply houses, are ideal for the purpose. A ring of Vaseline* should be thinly smeared on the inside of the dish just above the desired level of the liquid. The film of Vaseline will prevent the formation

---

* Vaseline is a protected name for petrolatum, a product of the Chesebrough Manufacturing Company.

of a crystalline crust on the sides of the dish. Without this protection the salt may creep up the inside of the dish, down the outside, and out over the top of the bench, thus becoming troublesome. The stock solution is then poured into the dish to the level of the film (half or three-quarters full). The solution may be decanted from the stock flask, but it is much better if it is filtered. Since crystals are formed spontaneously on dust particles, it is well to remove as much dust as possible by filtration, as well as minute crystal nuclei which may be floating in the stock solution. As the crystals grow and the solution evaporates, fresh solution may be filtered into the bath from time to time from the stock flask, which is kept beside the crystallizing dish. To prevent the showers of crystals from occurring in the bath (owing to supersaturation) it is convenient to dilute the bath slightly by adding several drops of water before handling the solutions.

3. *Care of the crystallizing bath.* Because of the increase of solubility of salts with temperature, the bath should be kept at as nearly constant temperature as possible. An easy way is to keep the bath in a closed basement room. The usual cellar responds sluggishly to changes in weather, and in such a place the bath can adjust to the slow change in temperature without detrimental effects on the crystal. The effect of slow changes in temperature will be seen in the appearance of clear and cloudy zones or "phantoms" in the crystal, but, if the changes are too violent, the crystal will not grow at all. Rounding of the upper edges of the crystal indicates a too sharp temperature rise, the warmed solution rising to the top of the dish having dissolved part of the crystal. After good temperature conditions have been assured, the rate of evaporation must be adjusted to a point where the crystal does not grow too fast and become ill-formed. Such adjustment is made by covering the dish with a watch glass, or larger dish, and varying the opening to the air with props until optimum conditions are obtained. The correct opening can be found only by experiment.

4. *Care of the growing crystal.* The fresh bath after a short time will yield several small seed crystals spontaneously. Two or three of the best of these are chosen, and all others are removed with chromium-plated forceps. If there are too many, it may be necessary to prepare a new bath and place the chosen seeds in it. If the evaporation rate is slow enough, these seeds will continue to

grow without the formation of any new seeds. As the crystal grows larger, satellite crystals, or "suckers," may become attached to it. These can be broken off and removed, and the large crystal will quickly heal itself. If a symmetrically developed crystal is desired, the crystal must be turned over on a new face every day.

Although the baths require little care, that care must be constant. They should be examined at least once every day. If they are neglected once, control of the crystal growth will be lost, and the crystals will be ruined. The period of growth may vary from a few weeks to several months. A crystal of copper sulfate weighing approximately one pound, grown in this way, required about five months.

The effect that impurities may have on crystal form is well illustrated by potassium ferricyanide. From pure solutions, this salt forms long slender pencils (Figs. 168a). If the solution contains 5 per cent of alkali (potassium hydroxide), the crystals become more equally developed (Fig. 168b), and much larger crystals can be grown. In growing these crystals, when the bath is replenished, the bath solution should be discarded, and the crystal immersed in an entirely fresh bath, in order to prevent building the alkali concentration too high. Solution impurities, as well as different physical conditions (temperature, pressure, etc.), can have a profound effect on the habit of crystals. We have here a clue to the reason for the wide variety of forms observed on crystals of such minerals as calcite.

Further experimentation with other salts will show a wide range of crystal stability and crystallizing power. Thus sodium nitrate ($NaNO_3$) and sodium chloride ($NaCl$) generally form skeletal crystals filled with liquid inclusions; potassium chromate ($K_2CrO_4$) crystallizes ordinarily in highly irregular and distorted shapes. Other salts, such as nickel sulfate ($NiSO_4 \cdot 6H_2O$), grow excellent large crystals, but they decompose slowly in the air (usually by loss of water of crystallization), become opaque, and may fall to a powder.

# CHAPTER IV

# *Physical Properties*
# *of Minerals*

In addition to the external form of minerals as shown in crystals and crystalline aggregates there are many other physical properties that characterize minerals and aid in their identification. Most of these properties can be divided into three groups, based on: (1) cohesive forces, (2) density, (3) the action of light. Other properties less frequently used in identification depend upon heat, electricity, magnetism, taste, and odor.

## 1. PROPERTIES DEPENDING ON COHESION

The properties of minerals that depend upon the atomic forces of cohesion include cleavage, fracture, hardness, tenacity, and elasticity.

## Cleavage

It has been repeatedly stated that not only does the crystal form reflect the atomic structure, but also the internal orderly arrangement of atoms is revealed by the *cleavage*. Mineral cleavage may be defined as the *natural easy fracture which yields more or less smooth surfaces in certain crystallographic directions*. A cleavage marks the direction in which the forces binding the atoms together are relatively weak. Thus galena is said to have *cubic cleavage* because it separates readily in a direction parallel to each pair of parallel cube faces. In other words, when fractured, galena breaks into a multitude of little cubes (Fig. 170). Fluorite has octahedral cleavage, since the separation is easy parallel to each pair of octahedron faces; hence from a single crystal we may with care break out an octahedron. This cleavage octa-hedron, by the way, is readily distinguished by careful examination from an octahedral *crystal* because the faces are uneven and splin-tery, not uniform like the normal faces of a crystal. Even when the faces of a crystal are rough and uneven, they are quite differ-ent from the surfaces formed by cleavage, however perfect.

The octahedral cleavage of fluorite is seen also in a cubic crystal, since its solid angles may be easily broken off, giving a form like Fig. 16 on p. 23, which we have learned is a cube modified, or with its angles replaced, by the planes of an octahedron. Again, from a piece of sphalerite a dodecahedron may be broken out because

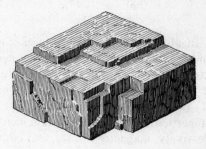

FIG. 170.  Cubic Cleav-age, Galena.

FIG. 171.  Rhombohedral Cleavage, Calcite.

of its perfect *dodecahedral cleavage*. If this is difficult because the fragment is too small, the skillful observer can prove that the cleavage faces have the position of dodecahedral planes and make angles of 60° and 120° with each other. Further, a piece of calcite breaks readily into a number of rhombohedrons, all of the same angle, and hence it is said to have perfect *rhombohedral cleavage* (Fig. 171). In the same way we find that amphibole has *pris-matic cleavage;* mica has highly perfect *basal cleavage* or cleavage parallel to the end or basal pinacoid, yielding excessively thin sheets. Topaz has also basal cleavage. Gypsum has perfect cleavage parallel to the side pinacoid of the crystal, yielding plates almost as thin as those of mica. These plates show two other cleavages on the edges, but differ from each other, as is more fully described under the description of this species. Feldspar shows two cleavages, both nearly perfect but one a little more so than the other, and these make an angle of 90° or nearly 90° with each other.

All these are examples of perfect cleavage, and it will be seen at once how important a property cleavage is.

But the cleavage is not always so perfect as in the examples given; in some minerals it is obtained with difficulty, or the surfaces yielded may be only partially smooth; in such minerals it is said to be imperfect, or interrupted, or difficult. Occasionally cleavage exists where it is so hard to obtain that it is ordinarily not noted at all. Thus crystallized quartz usually shows only a conchoidal fracture, and the absence of cleavage is a property that at once distinguishes it from the feldspar with which it is often associated. However, a crystal of quartz which after being heated has been plunged into cold water often shows cleavage parallel to the rhombohedral planes, and perhaps also parallel to the prism. In reporting cleavage one should give both the direction and the quality. For example, halite has perfect cubic cleavage, beryl has poor basal cleavage.

It is helpful to remember that, even if a crystal does not actually show broken surfaces, the cleavage is often clearly indicated by a fine pearly luster on the face of the crystal to which it is parallel. This is seen on the basal plane of apophyllite and on the side pinacoid of stilbite. This pearly luster is due to the presence of cleavage rifts, though the crystal has not actually parted, just as a pile of thin glass sheets or many layers of closely packed cellophane shows a pearly luster because of the repeated reflections. Similarly the cleavage rifts can often be seen in a transparent crystal; a flat clear crystal of barite or of celestite (Fig. 172) often shows the prismatic cleavage in two directions making an angle of about 104° with each other.

A massive specimen of a mineral may show cleavage as a multitude of little smooth faces changing position with that of the grains to which they belong; if these grains are very small, the cleavage may appear only as a fine spangling of the surface, as was mentioned on p. 59.

## Fracture

The nature of the surface given by fracture, when not the smooth surface of cleavage, is often an important property and may aid in distinguishing mineral species. Thus glass, as well as quartz and many other minerals, shows a shell-like fracture surface which is called *conchoidal* (Fig. 173) or, if less distinct, subconchoidal.

More commonly the fracture is simply said to be *uneven*, when the surface is rough and irregular. Occasionally it is *hackly*, like a piece of fractured iron. *Earthy* and *splintery* are other terms sometimes used and easily understood.

  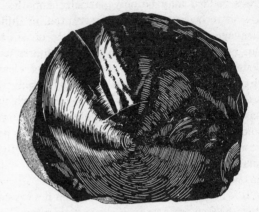

Fig. 172.  Cleavage in     Fig. 173.  Conchoidal Fracture in
Barite.                    Obsidian.

## Hardness and Tenacity

The resistance which the smooth surface of a mineral offers to a point or edge tending to scratch it is its *hardness* (designated by $H$). A diamond easily makes a scratch on a smooth topaz crystal; the topaz scratches a quartz crystal; the quartz scratches a glass surface; and the glass in turn scratches calcite. Each substance named is harder than that which it scratches, or, in other words, softer than the one by which it is scratched.

A number of common minerals have been selected for the comparison of hardness and are designated by the numbers from 1 to 10, as given in the accompanying list. Crystallized varieties should be taken in each case, that is, a crystal with even surfaces or a smooth cleavage fragment.

### Scale of Hardness

1. Talc.
2. Gypsum.
3. Calcite.
4. Fluorite.
5. Apatite.
6. Orthoclase.
7. Quartz.
8. Topaz.
9. Corundum.
10. Diamond.

When a mineral is said to have a hardness of 4 it means that it is scratched as easily as fluorite.  If the hardness of a mineral is given as 5½, the mineral is a little harder than apatite and a little less hard than orthoclase.

The student should practice with the minerals in this scale up to corundum, and then with them experiment upon some other known minerals, until he learns to what degree of hardness each of the numbers corresponds.  He should soon become so proficient as to be able to determine the lower grades of hardness by scratching with his knife (a hardness of about 5½) without the use of the reference minerals in the scale.

He will find at once the following general distinctions between minerals of the several grades:

No. 1 has a soft, greasy feel like talc and graphite, and flakes of it will be left on the fingers.

No. 2 can be scratched easily by the fingernail, as a cleavage piece of gypsum.

No. 3 can be easily cut by a knife and just scratched by a copper coin but is not scratched by the fingernail.

No. 4 is scratched by a knife without difficulty but is not so easily cut as calcite.

No. 5 is scratched with a knife with difficulty.

No. 6 is not scratched by a knife but is scratched with a file and will scratch ordinary glass.

No. 7 scratches glass easily but is scratched by topaz and a few other minerals.  Minerals that are as hard as, or harder than, quartz are few and include most of the highly prized gems.  Some further notes on hardness are given in Chapter VIII.

The beginner will need a word of advice in regard to testing for hardness.  In the first place, treat a mineral, especially a crystal, with as much consideration as possible.  A scratch on a piece of plate glass, like a daub of colored paint on a white wall, is a little thing, but it may have a sadly disfiguring effect.  Likewise, a scratch on a crystal disfigures it and destroys its value in large measure in the eyes of one who is a true mineralogist.  Hence make as minute a scratch as possible, not longer than this —, and, if possible, put the scratch where it will show least.  Treat the crystal, then, as if it had feelings that would be hurt by the cut, and never scratch its smooth surface wantonly.  If possible, identify a fine crystal by means of other tests.

In making the hardness test, it is necessary to be sure to distinguish between a real scratch on a smooth surface and the crushing of a rough surface by the knife-edge; a very hard mineral may often appear to be scratched in this way. The danger of making a mistake of this kind is less if, besides the useful knife-point test, the mineral is rubbed on a piece of glass. Have a small piece at hand (do not disfigure a windowpane). Do not make the opposite mistake of calling a white mark left by a soft mineral on the glass a scratch. The mark of a soft mineral can be rubbed off; a true scratch is permanent.

It is necessary to remember that minerals are often altered; that is, they may have undergone some chemical change, particularly on the surface, which has rendered them soft when the original mineral was really hard. Thus it is often easy to make a scratch on a crystal of corundum because of a surface change to mica, although the mass within is very hard. Make sure you are scratching a fresh, unaltered surface.

There are other properties also depending upon the force of cohesion acting between the atoms of a mineral. They include the following, which are grouped under the general head of *tenacity*.

**Malleability.** A mineral is malleable if it can be flattened under the blow of a hammer without breaking or crumbling into fragments. Malleability is conspicuously true of gold and silver and makes it possible to beat out gold into extremely thin leaves. The property of malleability belongs only to the native metals and, in an inferior degree, to a few compounds of silver.

**Ductility.** A substance that can be changed in shape by pressure, especially if it is capable of being drawn out into the form of wire, is ductile. Gold, silver, and copper are ductile. Among minerals only the native metals possess this property in a high degree.

**Sectility.** A mineral is sectile if it can be cut by a knife like cold wax, so that a shaving may be turned up with care, and yet the mineral breaks with a blow and is not completely malleable. Cerargyrite is eminently sectile. Gypsum and chalcocite are imperfectly sectile. No sharp line separates the minerals which show this property from the truly brittle minerals.

**Flexibility.** A mineral that bends easily and stays bent after the pressure is removed is flexible. Flexibility is shown by talc.

**Brittleness.**  A mineral is brittle if it separates into fragments with a blow of the hammer or with a cut by a knife.  Brittleness is observed, in varying degrees, in most minerals.

**Elasticity.**  When a mineral is capable of being bent or pulled out of shape and then returns to its original form when relieved, it is elastic.  Elasticity is shown well by a plate of mica.

## 2. SPECIFIC GRAVITY OR RELATIVE DENSITY

It has been shown that much can be learned about a mineral by visual inspection for, if crystals are present, their symmetry may enable one to place it at once in a crystal system.  If it is not in crystals, at least the habit of the aggregate and the presence or absence of cleavage can be observed.  By a simple trial its hardness may be determined.  While holding it one should note also whether it seems distinctly heavy or light as compared with common substances of similar appearance.  In this way one can obtain the first approximation of its density.

It is necessary at the outset to have a clear picture of what a difference in density means.  Suppose two balls are of the same size, but one is made of wood and the other made of iron.  The one of wood may be lifted easily, whereas the one of iron may be too heavy to move.  This example shows that in a piece of iron there is more matter to move—it has a greater mass than wood of the same bulk.  In other words, it has a greater density.  This is generally expressed by saying that the weight of a unit volume of iron is greater than that of a unit volume of wood.

We may say, consequently, that the relation of the densities of different bodies is given by the weights of blocks having the same bulk.  Suppose we could cut out similar blocks of aluminum and iron and weigh them.  The weights would be not far from the ratio of 1 to 3, and this would be the relation in density.  Now, to make this comparison for all bodies, it is evidently necessary to choose one of them as the standard and make the comparison with this.

The standard substance adopted is pure water, and, if great accuracy is required, it must be taken at the temperature 39.2° F (4° centigrade), the point at which it has its maximum density; for water that is growing cooler or warmer than 39.2° F expands a little and grows less dense.  The density of minerals is then compared with water, and this density is called the *specific gravity*. Consequently, the specific gravity of a mineral may be stated as the weight of a fragment divided by the weight of an equal volume of water. The specific gravity of sulfur is 2; of corundum, 4; of pyrite, 5, etc., these numbers mean that these minerals are respectively 2 times, 4 times, and 5 times as dense as water or, in other words, a given bulk—a cubic foot for example—weighs 2, 4, and 5 times more than the same bulk—a cubic foot—of water.

In order to find the specific gravity, it is not practicable to compare directly the weights of equal volumes because (though it is easy to weigh, for example, a piece of calcite) it is not possible to get the volume with sufficient accuracy. Hence it is necessary to make use of a well-known principle in hydrostatics, that when a body is immersed in water it is buoyed up to such an extent that it weighs less than it does in the air; and this loss of weight is equal to the weight of the water it displaces. Hence, if we find first the weight of the fragment on the pan of a delicate balance, and then its weight while immersed in water (it being suspended from the pan by a fine thread), and subtract the two weights, the difference is the weight of the equal volume of water.

For example, the weight of a little quartz crystal is 3.455 grams in the air; in water it is 2.156; the loss of weight, or weight of a volume of water exactly equal to it, is therefore 1.299; hence the specific gravity is:

$$\frac{3.455}{3.455 - 2.156} = \frac{3.455}{1.299} = 2.66$$

In the description of species the specific gravity is often expressed by the initial letter $G$; thus for quartz $G = 2.66$.

A spring balance (Fig. 174), like that of Jolly,* makes the operation very easy. This consists of a delicate brass spring, one end of which is attached to the top of a vertical scale; from the other end hang two pans, the lower one of which is immersed in water. The small fragment whose specific gravity is to be determined is placed in the upper pan, and the amount that the spring is stretched from the original zero position is noted by the number $(N_1)$. It is then placed in the lower pan, and the scale number $(N_2)$ again noted.

* The Jolly balance illustrated in Fig. 174, considerably improved over the original model, is manufactured by Eberbach & Son Company, Ann Arbor, Michigan.

The specific gravity is then given by the expression

$$G = \frac{N_1}{N_1 - N_2}$$

In this case we do not have the actual weights given, but, since specific gravity is a ratio, one can use numbers that are proportional to them.

A simple balance for determining the specific gravity, which also does away with the necessity of using definite weights, is shown in Fig. 175 (one-sixth natural size) and can easily be constructed with a little care. After the frame is made with the two uprights, a thin piece of wood is cut in the form of the steelyard beam, *abc*. This should be graduated into inches and tenths of inches or into centimeters and millimeters from *b*, the axis, to the end *c*; the distance from *b* to *a*, where the pans are supported, should be 4 inches. A fine wire passed through the beam at *b* serves as the axis, and the pans are also held by fine wires, preferably of platinum; the lower pan is immersed in water. A small piece of lead permanently placed between *b* and *a* serves to counterpoise the long arm, and a little rider is first moved to some point, as *d*, where it

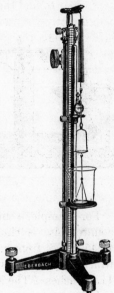

Fig. 174. Jolly Balance.

serves to make *bc* exactly horizontal, as shown by the mark on the upright near the end *c*. A number of weights are needed, which may be easily made of soft copper wire bent with a hook at one end, or they may be little glass tubes containing 1, 2, 3 or more shot and with a wire hook fused in at one end (see those at *g*). These weights need have no exact weight, but several varying from approximately ¼ to 2 grams are convenient. The mineral fragment* is first placed in the pan *e*, a suitable counterpoise is chosen, and the number ($N_1$) on the scale where it must be

---

* A fragment or crystal as heavy as 7 grams may be employed with the weights mentioned; but it should be understood that it is not necessary to know the actual weight either of the fragment or of the counterpoise.

placed to make *bc* again horizontal is recorded. It is then transferred to the pan *f*, immersed in the water, and the scale number
again noted ($N_2$), the same counterpoise being employed. The
first number then divided by the difference of the two numbers
gives the specific gravity:

$$G = \frac{N_1}{N_1 - N_2}$$

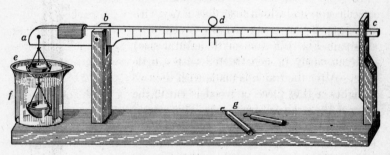

FIG. 175.   Beam Balance.

For example, a small pyrite crystal is placed in the pan *e*, and a
counterpoise is chosen which balances the beam when placed
14.45 inches from *b*; the crystal is then transferred to the pan *f*,
which requires that this same counterpoise be moved back to
11.55 inches.   The specific gravity is thus found to be

$$\frac{14.45}{14.45 - 11.55} = \frac{14.45}{2.9} = 4.98$$

It is evident that the accurate determination of the specific
gravity is a somewhat difficult matter, requiring a good deal of
care, but, as was suggested at the beginning, the hand can often
give very valuable information in this direction after a little
training.   We are accustomed to handle fragments of rocks, such
as granite (*G* about 2.7), marble (*G* = 2.7), sandstone and quartz
(*G* = 2.66), and like substances, and we know about what weight
to expect; thus, if we pick up a piece of barite (heavy spar),
perhaps thinking it is marble, which it may resemble so closely
that the unaided eye cannot distinguish between them, we notice
or should notice at once that it is unexpectedly heavy, for its specific
gravity is 4.5.   So a piece of corundum, we say, feels heavy because it has a specific gravity of 4.   This last, by the way, is an

interesting case, because corundum is the oxide of the very light
metal aluminum and its relatively high density is connected with
its great hardness; in other words, it is evident that its atoms must
be crowded together. In the same way we note that the substance
carbon forms the hard and relatively heavy* mineral diamond
($G = 3.5$) and the soft and light mineral graphite ($G = 2.2$). In
other words, the density depends not only upon the kinds of atoms
(all lead compounds, for example, have necessarily a high specific
gravity), but also upon the way in which they are packed together.

Again, if we take up a metallic mineral, we compare it, perhaps
without conscious thought, with other common metallic substances.
Hence a piece of aluminum seems very light because the specific
gravity is only 2.5, about one third that of iron and less than one
quarter that of silver. On the other hand, a fragment of galena
seems heavy because its specific gravity is 7.5 or nearly equal to
that of metallic iron.

As is stated on p. 81, all minerals can be divided into two groups:
(1) those that do not have a metallic appearance and (2) those
with a metallic appearance.

1. *Nonmetallic Minerals* may be roughly subdivided into three
classes:

   *a*. Minerals of relatively low density with specific gravity not
higher than 2.5. Examples are:

|         | $G$  |             | $G$  |
|---------|------|-------------|------|
| Borax   | 1.7  | Stilbite    | 2.2  |
| Sulfur  | 2.05 | Gypsum      | 2.3  |
| Halite  | 2.1  | Apophyllite | 2.4  |

The zeolites (as stilbite above) mostly fall between $G = 2.0$ and
$G = 2.3$.

   *b*. Minerals of average density; specific gravity 2.6 to 3.
Common examples are:

|           | $G$  |           | $G$       |
|-----------|------|-----------|-----------|
| Nepheline | 2.6  | Feldspar  | 2.6–2.75  |
| Quartz    | 2.66 | Talc      | 2.8       |
| Calcite   | 2.7  | Muscovite | 2.9       |
|           |      | Anhydrite | 2.9       |

* In these statements, as in some similar cases, the word *heavy* is used instead of *dense*, that is, of high specific gravity, and also the word *light* to express the opposite character; the meaning will be clear even if these terms are
not quite scientifically employed.

The common minerals tourmaline ($G = 3.0$–$3.2$), apatite ($G = 3.2$), idocrase ($G = 3.4$), amphibole ($G = 2.9$–$3.4$), pyroxene ($G = 3.2$–$3.6$), epidote ($G = 3.25$–$3.5$) fall between this and the following group. Some varieties of garnet also belong here; others have a specific gravity up to 4.3.

*c.* Minerals of high density: specific gravity 3.5 or above; a fragment seems heavy in the hand; if the specific gravity is above 4.5 it seems very heavy. Examples are:

| | G | | G |
|---|---|---|---|
| Topaz | 3.5 | Witherite | 4.3 |
| Diamond | 3.5 | Barite | 4.5 |
| Staurolite | 3.7 | Zircon | 4.7 |
| Strontianite | 3.7 | Scheelite | 6.0 |
| Celestite | 3.9 | Cassiterite | 7.0 |
| Corundum | 4.0 | Cinnabar | 8.1 |
| Rutile | 4.2 | | |

The compounds of lead also belong here, the commonest of which are: cerussite, the carbonate; anglesite, the sulfate; and pyromorphite, the phosphate. They all have a specific gravity between 6 and 7. Compounds of iron (such as siderite, $G = 3.8$), of copper (such as cuprite, $G = 6.0$), of silver (such as cerargyrite, $G = 5.5$), and of the other heavy metals also have high specific gravities.

*2. Minerals with Metallic Luster* have an average specific gravity of about 5, that of pyrite and hematite. If the density is much lower than 4, the mineral seems light in the hand, like graphite ($G = 2$). If the density is as high as 7, or above, it seems notably heavy, like galena ($G = 7.5$). Crystallized uraninite, which has a submetallic luster, has the remarkably high specific gravity of 9.7 when unaltered. The variation is wider with the metallic minerals than with those of nonmetallic luster, as will be seen by comparing the densities of the common metals.

DENSITIES OF THE COMMON METALS

| Magnesium | 1.8 | Copper | 8.9 |
|---|---|---|---|
| Aluminum | 2.5 | Bismuth | 9.8 |
| Arsenic | 5.7 | Silver | 10.6 |
| Antimony | 6.7 | Lead | 11.4 |
| Zinc | 7.1 | Mercury (liquid) | 13.6 |
| Tin | 7.3 | Gold | 19.3 |
| Iron | 7.8 | Platinum (pure) | 21.5 |

The quick judgment that comes with practice is always of value, but it should be applied with discretion, for the mineralogist must be continually on his guard lest he be misled.

In the first place, the size of the mass is an important factor, for a big lump of quartz seems heavy, of course, though its specific gravity is not relatively high. Also we may get a wrong impression in handling a specimen if the mineral we are interested in forms only a small part of it; a little galena in a large mass of quartz will not make it heavy. If the mineral is open and porous and is made up of interlacing fibers, like some specimens of cerussite, it may appear light, even if the specific gravity is actually high. The eye may thus be deceived by the appearance of bulk, if the solid mass present is not great. Some further suggestions on this subject are given in the closing chapter of this book.

## 3. PROPERTIES DEPENDING UPON LIGHT

Of the properties observed by the eye several have not been mentioned in detail as yet. These are the properties that depend upon the reflection or absorption of the light: (1) the *luster* or the appearance of the surface independent of the color, due to the way the light is reflected; (2) the *color;* and (3) the degree of *transparency*.

### Luster

The difference in luster is not always easy to describe, but the eye notes it at once and, after a little training, seldom makes a mistake.

The kinds of luster distinguished are as follows:

**Metallic:** the luster of a metallic surface like steel, lead, tin, copper, gold, etc. This is not always easy to distinguish, for the luster is not called metallic unless the mineral is quite opaque, so that no light passes through even very thin edges. The luster of some minerals, like columbite, is said to be submetallic when it lacks the full luster of the metals. A few minerals have varieties with metallic and others with nonmetallic luster; hematite is an example.

**Nonmetallic:** A number of types of nonmetallic luster represented by the following terms can be distinguished:

*Vitreous,* or glassy luster: that of a piece of broken glass. This is the luster of most quartz and of a large part of the nonmetallic minerals.

*Adamantine,* or the luster of the diamond. This is the brilliant, almost oily, luster shown by some very hard minerals, like diamond and corundum, and also by some compounds of the heavy metals, like the carbonate (cerussite) and the sulfate (anglesite) of lead. All these refract the light strongly or have a high refractive index.

*Resinous or waxy:* the luster of a piece of resin, as shown by most kinds of sphalerite; near this, but often quite distinct, is *greasy luster*, shown by some specimens of milky quartz and nepheline.

*Pearly,* or the luster of mother-of-pearl. This is common when a mineral has very perfect cleavage and hence has partially separated into thin plates. Thus the basal plane of crystals of apophyllite shows pearly luster.

*Silky,* the luster of a skein of silk or a piece of satin. This is characteristic of some minerals in fibrous aggregates, such as the variety of gypsum called satin spar, and of most asbestos.

The luster of minerals is described also according to the brightness of the surface; it is called *splendent* in freshly fractured galena, but *dull* in jasper. Other surfaces may be described in self-explanatory terms such as *glistening* or *glimmering* according to the nature of the surface.

## Color

The variation in the color of minerals is great and includes all hues from white to black, running through many shades of red, yellow, green, and blue. Most of the terms used in describing the color are so familiar that they explain themselves. Thus we speak of azure-blue, cherry-red, and so on.

Color, however, varies not only from one species to another but also within a single species, where there may be a wide range. It is, therefore, very important for the mineralogist to learn in which minerals color is a constant property and can be relied upon as a distinguishing criterion. As we study the different minerals we shall see that those with metallic luster vary little in color but that many with a nonmetallic luster vary widely.

In general, therefore, the color of minerals with metallic luster is a very important and constant character, and such variations as are noted are due chiefly to a slight change of the surface. Variations occur if the mineral becomes tarnished; that is, a mineral may be dull in some specimens, or perhaps bright-colored in others. Thus galena has a very characteristic bright bluish lead-gray color, which may become dull if the surface has been long exposed to the air. The pale brass-yellow of pyrite and the still paler shade of the same color characteristic of fresh marcasite are readily distinguished, but both minerals, particularly marcasite, are subject to change by tarnishing; tarnishing may make the distinction between the two impossible if no fresh surfaces are visible.

Hence it is necessary to be sure that what we see is the true color of a mineral, that is, the color obtained from a fresh surface by fracture. Another striking example is given by bornite, which always has a peculiar reddish bronze color on a fresh surface. Bornite changes or tarnishes so easily that it almost immediately becomes colored bluish or purple, the color depending on the length of time it has been exposed. Hence the mineral is sometimes called purple copper ore or variegated copper ore. So too chalcopyrite, which is bright brass-yellow when fresh, may become iridescent on exposure.

Of the minerals with nonmetallic luster a few always have the same color. Thus malachite is green, azurite is blue, rhodonite is red. In this group of minerals, however, different specimens of a single species show more often a range of colors. For example, tourmaline may be colorless but other varieties are different shades of red, blue, green, brown, and black. Moreover, tourmaline may show several different colors within the same crystal. Likewise corundum is found in many different colors. The name *ruby* has been given to the clear red variety and the name sapphire to the clear blue. Here, as in many other minerals, it is difficult to tell precisely what is the slight chemical change upon which the color depends.

**Streak.** It is often important, especially with a mineral having metallic luster, to test the color of the fine powder, or the color of the streak. The slight scratch which is given to test the hardness will often show the streak, but a better way is to have at hand a piece of unglazed white procelain, or ground glass, upon which the mineral can be rubbed. This method shows, for example, that an iron-black hematite crystal with a bright metallic luster has a red streak. For many minerals this is a highly diagnostic property.

Minerals of nonmetallic luster usually have a streak which

differs but little from white even if the mineral itself is dark-colored or even black, as, for example, the different varieties of tourmaline.

## Transparency

A mineral is said to be *transparent* when it is so clear that an object can be seen through it with perfect distinctness, as through a piece of window glass, a plate of selenite, or a thin sheet of mica.

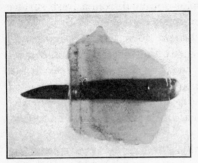

Fig. 176. Transparent Mica.       Fig. 177.   Translucent Quartz.

A mineral is *translucent* when it transmits light, as a piece of thin porcelain does, and allows only the outline of an object to be seen through it.   It is *subtranslucent* when light is transmitted only on the edges.

If no light is transmitted even on thin edges the mineral is said to be *opaque*.   Most metallic minerals are opaque.

## Play of Colors

When describing the color of a mineral, some peculiarities in its distribution may be noted and may receive special names.   A mineral is said to show a play of colors when, like precious opal, it exhibits internally the various prismatic colors when the mineral is turned.

## Opalescence

A pearly reflection from the interior of a mineral, like the effect of a glass of water to which a few drops of milk have been added, is called opalescence because such an effect is common with opal. Opalescence is also seen in moonstone.

## Iridescence

A mineral is iridescent when it shows a series of prismatic colors either on the surface or in the interior. External iridescence is caused by the presence of a thin surface film or coating; internal iridescence is often caused by the presence of thin twinning lamellae or minute air spaces, along cleavage planes.

Fig 178.   Asterism in Phlogopite.

## Asterism

Some minerals show a peculiar starlike effect known as asterism, when they are viewed by reflected light, as is well shown in some quartz and in those varieties of corundum known as star sapphire and star ruby. A similar effect is seen when a point source of light is viewed through certain types of phlogopite mica (Fig. 178). In both cases this phenomenon is caused by peculiarities of structure or inclusions oriented along crystal directions. Both types of asterism give a six-pointed star dependent on the symmetry of the crystals.

## Fluorescence and Phosphorescence

One of the most fascinating properties of certain minerals, and one that first interests many in the study of minerals, is the phenomenon of *fluorescence*.

We are all familiar with the spectral colors from red to violet. Of these red has the longest wavelength and violet the shortest. Beyond the violet which we can see are still shorter wavelengths known as ultra-violet. If, on exposure to this ultra-violet light in a dark room, a mineral emits visible light it is said to fluoresce. If one examines his mineral collection under such conditions, he may make some remarkable discoveries. A drab piece of calcite may become a vivid pink, and a piece of chalky white scapolite may turn a bright yellow, whereas the most colorful crystal in the collection as seen in white light may have no color at all.

Unlike most of the other properties of minerals, fluorescence is difficult to predict. If, for example, we are sure a specimen is calcite, we can say it will have a hardness of 3, a specific gravity of about 2.72, and a good rhombohedral cleavage; but the only way in which we can be sure that it will fluoresce is to try it. Two specimens of calcite may look identical; yet one will fluoresce, and the other will not. The name of this property comes from *fluorite*, because it was first observed in this mineral. However, only a few of the fluorite specimens will fluoresce. Fluorescence, therefore, cannot be used as a tool in determinative mineralogy except in rare cases. One such exception is willemite from Franklin, New Jersey, which fluoresces a bright green. Another is the mineral scheelite which usually fluoresces a pale blue.

With the increasing interest in the study of fluorescent substances, new and better sources of ultra-violet light are each year being made available. The student of mineralogy can obtain inexpensive argon-filled bulbs or "black lights" that screw into an ordinary light socket for the first examination of his specimens. A number of somewhat more expensive lights can be obtained that will excite a stronger fluorescence.

Just as the wavelength of visible light varies from place to place in the spectrum, so does the wavelength of ultra-violet light vary. One source of ultra-violet may have no effect on a specimen, whereas a light of different wavelength may produce a vivid fluorescence. If possible, therefore, one should have more than one source.

Advertisements of ultra-violet light sources will be found on the pages of *The Mineralogist* and *Rocks and Minerals,* two journals published for the amateur mineralogist.

Some fluorescent minerals will continue to glow after the ultra-violet light has been turned off. This property is known as *phosphorescence.* Only a few of the fluorescent minerals such as certain specimens of willemite and calcite show this property to a marked degree. Others may phosphoresce but for so short a period that the eyes cannot become adjusted to the darkness quickly enough to observe it.

### Thermoluminescence and Triboluminescence

Some minerals when heated below red heat will emit visible light and are said to be *thermoluminescent.* As with fluorescence, the color of the light emitted is quite independent of the color of the mineral. It is an interesting experiment to heat a thermoluminescent mineral in a darkened room. As the specimen begins to get warm, it will give off a faint glow that becomes brighter with continued heating. Some specimens will continue to glow for many minutes after removal from the flame. This property is most commonly observed in fluorite, and the variety known as *chlorophane* was so named because of the green light emitted. Other thermoluminescent minerals are scapolite, apatite, calcite, and feldspar; but again it should be pointed out that only a few specimens of a given mineral will have this property.

Some minerals when rubbed or struck with a hammer will emit light and are *triboluminescent.* This property is less spectacular than fluorescence and thermoluminescence and is usually observed only momentarily. However, it is easily seen in a dark room if two pieces of milky quartz are rubbed briskly together. Other minerals that commonly show this property are fluorite, sphalerite, and lepidolite.

### Optical Properties

Only a few words can be devoted here to the large and important group of optical properties of minerals. These properties depend upon the action of light as determined by the structure of the minerals. The understanding of this part of physical mineralogy depends first on a good knowledge of crystallography, and second on a mastery of optics, especially as pertains to the subject of

polarization.  Only two points can be touched upon here.  One
of these is double refraction, or the separation of a ray of light
passing through certain crystalline substances into two rays.  This
is indeed true of all transparent crystals, except those of the iso-

FIG. 179.   Double Refraction in Calcite.

metric system.  The only common mineral in which it is noted to
a marked degree is calcite, especially in the transparent variety
called Iceland spar.  Figure 179 illustrates this property well;
here the single cross on the paper beneath appears double to the
eye;  one cross (to the eye looking perpendicularly down on the
surface) has its arms in the continuation of the lines beneath, the
other is pushed to one side.  Neither cross appears quite black
except at the two points where they intersect.

The appearance of different colors when a crystal is viewed in
transmitted light in different directions is known as *pleochroism*.
If only two directions have distinct colors, the property is called
*dichroism*.   It is due to varying degrees of light absorption or the
differential absorption of various wavelengths of light.  Pleochroism is often seen in transparent crystals of epidote and in the
pink gem variety of spodumene, kunzite.

## 4. PROPERTIES DEPENDING UPON HEAT

The fusibility of minerals or the ease with which they can be melted is an important property of many minerals. This is discussed on p. 106 with the description and the use of the blowpipe.

Fig. 180. Lodestone showing Magnetism.

The heat conductivity of crystals is another property which has been measured for many crystals. As would be expected, it depends upon their crystal structure in different directions, but this and related subjects belong to advanced mineralogy.

## 5. MAGNETIC PROPERTIES

A few minerals have the property of being attracted by an ordinary horseshoe magnet. This is true of magnetite, the magnetic oxide of iron; of pyrrhotite, magnetic pyrites; and of some specimens of native platinum.

Some specimens of magnetite are themselves magnets and have the power of attracting little particles of iron or steel (Fig. 180). They have north and south poles and, if hung by a thread, will swing around until the poles come into the magnetic meridian, that is, the direction assumed by a compass needle. This kind of magnetite is called the lodestone. Pyrrhotite is much less magnetic than magnetite, and the magnetic varieties of platinum are not common; both may have polarity like the lodestone. A few minerals, such as hematite, ilmenite, and franklinite, are sometimes slightly magnetic but probably only because they contain a little admixed magnetite.

Many other minerals are attracted to a strong electromagnet, but they are not ordinarily thought of as magnetic minerals.

## 6. ELECTRICAL PROPERTIES

There are a number of electrical properties of minerals, but they belong to a more minute study of minerals and need be mentioned only briefly here.

A number of minerals, like sulfur, diamond, and topaz, take on a rather strong electric charge when rubbed, as with a piece of silk, and show this by their power of attracting light substances such as bits of straw or paper.

Another kind of electrical charge can be developed in certain minerals. For example, if pressure is exerted along an *a* axis of quartz, a positive electrical charge is set up at one end of the axis, and a negative charge at the other end. This is known as *piezoelectricity*. When similar charges are induced by carefully heating the crystal, the property is called *pyroelectricity*. These properties, particularly piezoelectricity, vary widely from one mineral to another, and the ones developing the larger charges are said to be more "active." Only those minerals that crystallize in a symmetry class that lacks a symmetry center can develop pyro- or piezoelectricity. In recent years many applications of piezoelectricity have been developed. The outstanding example is the use of quartz plates for the control of radio frequencies. Tourmaline also has been used to a lesser extent for similar purposes.

## 7. TASTE AND ODOR

Taste belongs only to the few minerals that dissolve to some extent in water. The terms employed are familiar and hardly need explanation. Saline means the taste of common salt; alkaline, of soda; bitter, of epsom salts; sour, of an acid; astringent, of iron vitriol; sweetish astringent, of alum; cooling, of saltpeter. Halite, ordinary salt, and sylvite can usually be identified by their taste.

An odor is characteristic of only a few minerals. Some varieties of limestone, barite, or quartz have a fetid odor, or odor of rotten eggs, especially when they are freshly broken or rubbed sharply; this is usually due to the presence of some sulfur compound. Moistened clay and some claylike minerals when breathed upon give off a peculiar argillaceous odor. Bitumen and some allied substances have a bituminous odor.

A sharp blow across the surface of a piece of arsenopyrite often produces a peculiar garlic odor, like that obtained by heating the same mineral on charcoal; it is, in fact, due to the same cause. Similarly a blow on a mass of pyrite may yield a sulfurous odor.

# CHAPTER V

# *The Chemical Properties*
# *of Minerals*

It has already been stated that every mineral is necessarily a chemical element or compound, and that this is the most essential point in the definition of a mineral. But to understand what a chemical compound is, and what relations different compounds bear to each other, requires some knowledge of the fundamental principles of chemistry. In the first place, it is necessary to understand what the chemical elements are.

**Chemical Elements.** One of the duties of the chemist in his laboratory is to analyze different substances, or in other words to separate them into the various kinds of matter which they contain. However, this process of analysis, or chemical separation, can be carried only so far, for the chemist soon obtains substances which he is unable to resolve further. Thus, if he takes a piece of calcite, it is easy by simply heating it to separate it into a white powder called lime and a gas called carbon dioxide. Then further, if the proper procedure is followed, the lime can be separated into the metal calcium, and a gas, oxygen; the carbon dioxide can be separated into the familiar substance carbon and the same gas, oxygen. But these three substances, calcium, carbon, oxygen, cannot be separated into simpler substances by ordinary chemical means; hence they are called elementary substances or elements. Again, common salt can be separated into two kinds of matter, the metal sodium and the gas chlorine, but neither of these can be separated further, and hence they are also called elements. So, too, galena can be separated into its elements, lead and sulfur.

These substances, then, into which all matter can be separated but which the chemist is unable to decompose further, are the *chemical elements*.

Now the chemist finds that, although there is no limit to the different kinds of bodies which he may be asked to analyze or separate into their parts, still they contain but a small number of distinct kinds of matter. If we consider only those that are commonly present, they are very few indeed. There are, it is true, ninety-two elements recognized by the chemist (ninety-four if we add the newly discovered neptunium and plutonium), but many of them are rare, and those that make up the chief part of the common minerals are hardly more than twenty.

The table on p. 94 gives the names of all the common elements and most of the rarer ones. With the names are given also the letter or letters by which they are represented in the kind of shorthand that the chemist employs; these letters are called the symbols of the elements. Thus oxygen is represented by the capital letter O; hydrogen by H; nitrogen by N; calcium by Ca; and so on. In many cases the initial letters of the Latin name of a metal are used, as Fe, from the Latin *ferrum*, for iron; Ag, from *argentum*, the Latin name of silver; Au, from *aurum*, gold; Sb, from *stibium*, antimony, etc.

**Atomic Weight.** The numbers placed after each symbol give the atomic weight of the element. We have spoken particularly of the minute particles or atoms of which the physicist tells us that matter is composed. We have also seen that the regular form of a crystal is due to the arrangement of these atoms as they are marshaled into place by the attractive forces acting between them when the solid is formed. It is the relative weight of the atom of each substance compared with a standard that is called its atomic weight. In determining these relative weights the chemist uses oxygen with an atomic weight of 16 as his starting point. Thus, an atom of hydrogen is found to have a weight of 1.008, and carbon, 12.010. It does not mean that the chemist actually weighs the atoms themselves but he compares the weights of two masses, for instance of oxygen and hydrogen, under such conditions that he is sure that he is comparing the same number of atoms, and hence he obtains the relative masses of the atoms.

It is found invariably true that, when different elements unite to form a certain compound, there is always a definite relation between the amounts by weight of each element and that these

## ATOMIC WEIGHTS
### 1947

| | Symbol | Atomic Weight | | Symbol | Atomic Weight |
|---|---|---|---|---|---|
| Aluminum | Al | 26.97 | Molybdenum | Mo | 95.95 |
| Antimony | Sb | 121.76 | Neodymium | Nd | 144.27 |
| Argon | A | 39.944 | Neon | Ne | 20.183 |
| Arsenic | As | 74.91 | Nickel | Ni | 58.69 |
| Barium | Ba | 137.36 | Nitrogen | N | 14.008 |
| Beryllium | Be | 9.02 | Osmium | Os | 190.2 |
| Bismuth | Bi | 209.00 | Oxygen | O | 16.0000 |
| Boron | B | 10.82 | Palladium | Pd | 106.7 |
| Bromine | Br | 79.916 | Phosphorus | P | 30.98 |
| Cadmium | Cd | 112.41 | Platinum | Pt | 195.23 |
| Calcium | Ca | 40.08 | Potassium | K | 39.096 |
| Carbon | C | 12.010 | Praseodymium | Pr | 140.92 |
| Cerium | Ce | 140.13 | Protactinium | Pa | 231 |
| Cesium | Cs | 132.91 | Radium | Ra | 226.05 |
| Chlorine | Cl | 35.457 | Radon | Rn | 222 |
| Chromium | Cr | 52.01 | Rhenium | Re | 186.31 |
| Cobalt | Co | 58.94 | Rhodium | Rh | 102.91 |
| Columbium | Cb | 92.91 | Rubidium | Rb | 85.48 |
| Copper | Cu | 63.54 | Ruthenium | Ru | 101.7 |
| Dysprosium | Dy | 162.46 | Samarium | Sm | 150.43 |
| Erbium | Er | 167.2 | Scandium | Sc | 45.10 |
| Europium | Eu | 152.0 | Selenium | Se | 78.96 |
| Fluorine | F | 19.00 | Silicon | Si | 28.06 |
| Gadolinium | Gd | 156.9 | Silver | Ag | 107.880 |
| Gallium | Ga | 69.72 | Sodium | Na | 22.997 |
| Germanium | Ge | 72.60 | Strontium | Sr | 87.63 |
| Gold | Au | 197.2 | Sulfur | S | 32.066 |
| Hafnium | Hf | 178.6 | Tantalum | Ta | 180.88 |
| Helium | He | 4.003 | Tellurium | Te | 127.61 |
| Holmium | Ho | 164.94 | Terbium | Tb | 159.2 |
| Hydrogen | H | 1.0080 | Thallium | Tl | 204.39 |
| Indium | In | 114.76 | Thorium | Th | 232.12 |
| Iodine | I | 126.92 | Thulium | Tm | 169.4 |
| Iridium | Ir | 193.1 | Tin | Sn | 118.70 |
| Iron | Fe | 55.85 | Titanium | Ti | 47.90 |
| Krypton | Kr | 83.7 | Tungsten | W | 183.92 |
| Lanthanum | La | 138.92 | Uranium | U | 238.07 |
| Lead | Pb | 207.21 | Vanadium | V | 50.95 |
| Lithium | Li | 6.940 | Xenon | Xe | 131.3 |
| Lutecium | Lu | 174.99 | Ytterbium | Yb | 173.04 |
| Magnesium | Mg | 24.32 | Yttrium | Y | 88.92 |
| Manganese | Mn | 54.93 | Zinc | Zn | 65.38 |
| Mercury | Hg | 200.61 | Zirconium | Zr | 91.22 |

weights are either the atomic weights or simple multiples of them, as given by the number of atoms present.

Just what this means will be shown by some examples. It was stated above that the chemist could break down common salt into sodium and chlorine. Now in doing this it is possible to find the weight of each. For instance, in 100 parts, the result is:

| | |
|---|---|
| Sodium | 39.32 |
| Chlorine | 60.68 |
| | 100.00 |

But the numbers 39.32 and 60.68 are in the ratio of 23 to 35.5 (39.32 : 60.68 = 23 : 35.5), which have been independently found to be the atomic weights of these two elements, and hence it is evident that for every atom of sodium there is one atom of chlorine. The brief expression for the composition of sodium chloride, or the formula as it is called, is NaCl, for Na is the symbol for sodium and Cl of chlorine.

The formula of a compound, therefore, gives simply the kinds of elements present, represented by their symbols, with small numbers written below, to show how many parts of each, that is, how many atoms are present in one molecule.

Calcium, also, unites with chlorine; and the compound calcium chloride, analyzed by the chemist, gives:

| | |
|---|---|
| Calcium | 36.04 |
| Chlorine | 63.96 |
| | 100.00 |

Here the numbers 36.04 : 63.96, expressing the ratio by weight of the two substances, are in the ratio of 40 : 71 or 40 : 2 × 35.5; hence a molecule of this compound contains one atom of calcium and two of chlorine, and the formula for it is $CaCl_2$.

Two other examples are gold chloride and tin chloride, analyzed with the following results:

| | | | | |
|---|---|---|---|---|
| Gold | 64.90 | | Tin | 45.47 |
| Chlorine | 35.10 | | Chlorine | 54.53 |
| | 100.00 | | | 100.00 |

For the gold chloride the ratio of 64.90 : 35.10 is as 197 : 106.5 or 197 : 3 × 35.5; hence the formula is written $AuCl_3$. Similarly for tin chloride the ratio of 45.47 : 54.53 is as 118.7 : 142 or 118.7 to 4 × 35.5; hence the formula is $SnCl_4$.

These examples illustrate the fact that the atomic weights of the given elements multiplied by the number of atoms gives the amount of each element present in a molecule of a given compound. They also show another important point: one atom of chlorine unites with one atom of sodium, but two atoms of chlorine with one of calcium, three with one of gold, and four with one of tin.

**Valence.** For hydrochloric acid the formula is HCl; in other words, one atom of hydrogen is present for one of chlorine. Water, however, has the formula $H_2O$, or it contains in a molecule two atoms of hydrogen and one of oxygen, and ammonia with the formula $H_3N$ has three atoms of hydrogen for one of nitrogen. It can be seen, therefore, that the various elements differ in the number of hydrogen atoms with which they will combine. This number, known as the *valence* of an element, determines how many atoms of another kind it can hold in combination. Thus the valence of sodium and chlorine is one, of calcium and oxygen two, of gold and nitrogen three, and of tin four.

There are many examples in chemical compounds and some among minerals in which we find two given elements forming more than one compound. In such cases at least one of the elements has more than one valence. For example, the chemist knows of one compound, FeO, another, $Fe_2O_3$, in which oxygen has, as always, a valence of two but the valence of iron is two and three respectively. Much attention must be given to this subject before it can be thoroughly understood, and for this study the student must have a good course in chemistry, including not only the study of some standard book but also practical work in the laboratory. But the explanations given should suffice to make it clear, first, what the elements are; second, what is meant by their atomic weights; and third, the significance of their combining power or valence.

The distinction between a chemical compound and a simple mixture of two elements is well illustrated by the air we breathe. The chemist finds by analysis that the air is nearly constant in composition, containing essentially in 100 parts, 78 parts by weight of nitrogen, 21 of oxygen, and 1 of argon. A variable amount of water vapor is also present and a small amount of carbon dioxide.

Is the air a chemical compound? The answer is given at once that it is not, for the simple reason (and there are others equally conclusive) that the ratio of 78 to 21 is not that of the atomic weights of the two principal elements present, namely 14 : 16, nor of any simple multiples of these. There are indeed several compounds of nitrogen and oxygen known to the chemist, namely, NO, $N_2O$, and $N_2O_3$, but air is merely a mixture of these two elements and is not chemically combined any more than is a mixture of salt and sand.

**Positive and Negative Elements.** One further point must be mentioned in regard to the compounds taken for illustration:

| | |
|---|---|
| Sodium chloride, NaCl | Gold chloride, $AuCl_3$ |
| Calcium chloride, $CaCl_3$ | Tin chloride, $SnCl_4$ |

The first element in these and similar formulas is a metal and the second a nonmetal; the first is said to be the positive element, the second is the negative element. Why the terms positive and negative are introduced is known to the student of electricity, for he has learned that, in the decomposition of a compound by an electric current, one element always goes to the positive pole or electrode, the other to the negative; the former is hence called the negative element, its atoms being attracted by the oppositely charged positive electrode, and the second the positive, since its atoms are attracted by the negative electrode. In this way it could be shown that the metals are positive in nearly all their compounds, whereas the nonmetals are negative. The elements known as the semi-metals are positive in some compounds, and negative in others. The positive element is always written first, as in the examples given above.

Arsenic is a semi-metal; in $As_2O_3$ and $As_2S_3$ it is positive, but in $FeAs_2$ and $CoAs_3$ it is negative. Similarly among the metals there are some which in compounds with certain elements (such as oxygen and sulfur) are always positive but when combined with certain other metals, may be negative.

The names given to the different chemical compounds are usually easy to learn and understand. In the description of minerals, in the pages that follow, both the chemical names and the formulas are given so as to familiarize the student with each method.

The following are a few special names given to certain oxides with which it is desirable to be familiar:

$Na_2O$, soda, instead of sodium oxide.
$K_2O$, potash, instead of potassium oxide.
CaO, lime, instead of calcium oxide.

MgO, magnesia, instead of magnesium oxide.
$Al_2O_3$, alumina, instead of aluminum trioxide.
$SiO_2$, silica, instead of silicon dioxide.

Since the classification of the minerals (see p. 123) described later
in this book is based on chemical composition, it is important that
the student be familiar with the common kinds of compounds
found among minerals.  The classification groups the minerals
into the following chemical classes: (1) *native elements*, that is, elements occurring uncombined with others; (2) *sulfides*, composed
of a metal and sulfur; (3) *oxides*, compounds composed of a
metal in combination with oxygen; (4) *halides*, compounds in
which a metal is combined with chlorine, fluorine, bromine, or
iodine; (5) *carbonates*, salts of carbonic acid having the radical
$CO_2$ in their formulas such as calcite, $CaO \cdot CO_2 = (CaCO_3)$;
(6) *phosphates*, those compounds in which oxides of phosphorus
are combined with a metal; (7) *sulfates*, salts of sulfuric acid in
which oxides of sulfur are combined with a metal; (8) *silicates*,
those compounds in which one or more metals are combined with
silica.

**Percentage Composition.**  It was shown on p. 95 that, from
the percentage of the different elements given in a chemical analysis
of a mineral, the chemical formula could be deduced with the aid
of the table of atomic weights.   Conversely, if the formula is given,
the *percentage composition*, or the amount by weight of each element
or group of elements present in the compound in one hundred parts,
can be easily calculated.   Thus sodium chloride has the formula
NaCl, and hence, if the atomic weights of sodium and chlorine are
taken, the relation of the weight of sodium to that of chlorine is as
23 : 35.5.   Now adding 23 and 35.5, we obtain 58.5, which is called
the weight of the molecule, or molecular weight.   Further, if
58.5 parts of the compound contain 23 parts of sodium, 100 will
contain 39.32 parts:

$$58.5 : 23 = 100 : 39.32, \quad \text{or} \quad \frac{23 \times 100}{58.5} = 39.32$$

and for chlorine:

$$58.5 : 35.5 = 100 : 60.68, \quad \text{or} \quad \frac{35.5 \times 100}{58.5} = 60.68$$

The percentage composition of sodium chloride is, therefore:

$$
\begin{array}{ll}
\text{Na} & 39.32 \\
\text{Cl} & \underline{60.68} \\
& 100.00
\end{array}
$$

Again, the formula of stibnite, $Sb_2S_3$, means that two atoms of antimony (Sb) unite with three of sulfur (S). But the atomic weights of antimony and of sulfur are 121.8 and 32 respectively. The molecular weight is, therefore, equal to

$$2 \times 121.8 + 3 \times 32 = 243.6 + 96 = 339.6$$

Hence, in 339.6 parts, 243.6 are antimony and 96 sulfur, and to find the amount of antimony in 100 parts we have the proportion:

$$339.6 : 243.6 = 100 : 71.74 \quad \text{or} \quad \frac{243.6 \times 100}{339.6} = 71.74$$

and for the amount of sulfur:

$$339.6 : 96 = 100 : 28.26 \quad \text{or} \quad \frac{96 \times 100}{339.6} = 28.26$$

The percentage composition is, therefore:

$$
\begin{array}{ll}
\text{Sb} & 71.74 \\
\text{S} & \underline{28.26} \\
& 100.00
\end{array}
$$

**Crystal Chemistry.** The study of the interrelation of chemical compounds and their crystal structure and form is a broad branch of the science that can be only touched upon here. Certain minerals have the same chemical composition as others, yet are given different names, and each has its own set of properties. Such compounds are said to be *dimorphous* if they exist in two modifications and *polymorphous* if they exist in more than two. The best example is carbon in the form of graphite and diamond. Another example is calcium carbonate existing as both calcite and aragonite.

Some minerals that have similar crystal structure but different chemical composition can crystallize together in varying proportions to form mixed crystals. This crystallizing together is known as *isomorphism*, and the intermediate compounds are said to form an *isomorphous series*. The outstanding example is in the plagioclase feldspars. Here albite and anorthite mix together in all proportions to form feldspars of intermediate compositions.

# CHAPTER VI

# *The Use*
# *of the Blowpipe*

As has already been explained, most minerals are chemical compounds, and the chemical composition of a given species is the most important of its many properties. Thus, when all other methods of identification fail, a chemical analysis must be made. To be able to tell the percentage of each element in a mineral we must have a quantitative analysis that can be carried out only by the skilled chemist.

However, in most cases where the identification is in doubt, a qualitative analysis would be all that is necessary. That is, we must determine what elements are present but without regard to their relative amounts. Since many of these chemical tests are laborious and require a considerable knowledge of chemistry, the mineralogist has devised for most of the common elements, simple and quick tests that are usually grouped under the heading *blowpipe tests*.

The results of the blowpipe analysis, together with the study of the physical properties of a given specimen, usually enable a mineralogist with a fair amount of experience to identify it, even if at first it was entirely unknown.

## DESCRIPTION OF APPARATUS

The following list includes the articles that are most essential for blowpipe work:

1. Lamp.
2. Blowpipe.
3. Platinum- or Nichrome-tipped forceps.

4. Charcoal.
5. Platinum wire.
6. Glass tubes.

1. *Lamp.* The most convenient form of lamp is a small Bunsen gas burner similar to that used by the dentist (Fig. 181); it is provided with a special jet (*e* in the figure). The burner can be connected with any ordinary gas outlet by a rubber tube so that it may be placed on the table in a position convenient for use. In the Bunsen burner, when the jet *e* is not inserted, the gas mingles

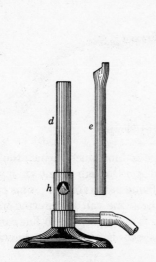

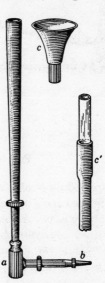

FIG. 181. Bunsen Burner.    FIG. 182. Blowpipe.

in the tube with the air that enters at *h*, and they burn together at the top in a very hot blue flame. Instead of the Bunsen burner an alcohol lamp may be employed; if neither of these is available, an ordinary candle is satisfactory.

When the jet *e* is inserted in the tube of the Bunsen burner, the air supply from the openings below is cut off, and the gas burns at the top with a yellow flame flattened by the shape of the jet; the convenient flame for ordinary use is about 1½ inches in height. This flame, about the size of an ordinary candle flame, is to be used with the blowpipe.

2. *Blowpipe.* A common form of blowpipe is shown in Fig. 182. It may be simple and inexpensive but should have an air chamber (*a*) to collect the condensed moisture from the breath. A separate tip (*b*), either of brass or platinum, with a fine hole, is often used,

but it is not absolutely necessary. It is essential that the hole, whether it is in the tip or in the tube itself, should be round and true, so that a moderate pressure of air will suffice to blow a clear blue flame. (See Fig. 186.) A trumpet-shaped mouthpiece (*c*) is usually furnished, but some prefer to dispense with it.

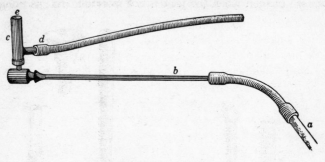

FIG. 183.   Improved Blowpipe.

Figure 183 illustrates another type of blowpipe in which the gas entering the tube *d* and air blown in at *a* through tube *b* are mixed in the fitting at *c*. A hot blowpipe flame issues from *e*. This type of blowpipe is advantageous, for the flame can be directed more easily. Moreover, if compressed air is available, the air line can be connected to *b* (Fig. 183) and the operator is relieved from blowing.

FIG. 184.   Forceps.

3. *Forceps.* A pair of steel forceps (Fig. 184) is needed, one end of which has platinum or Nichrome points; the forceps are self-closing by means of a spring, so that the piece of mineral to be heated, when it is placed between them, is firmly supported. At the other end are ordinary forceps for picking up small fragments; this end should never be inserted in the flame.

4. *Charcoal.* Several pieces of charcoal are needed. The most convenient are rectangular (Fig. 188) and about 4 inches long, an inch wide, and ¾ inch thick. The charcoal must burn without snapping and must leave very little white ash; with care one piece will last for many experiments if the surface is rubbed clean after each use.

5. *Platinum Wire.* A few inches of platinum wire, of the size designated as No. 27, are needed; directions for its use are given on p. 109.

6. *Glass Tubes.* Tubes of rather hard glass are required; it is convenient to have two sizes, with internal diameters of 4 and 6 millimeters respectively, but one size will suffice. The larger size should be cut into pieces about 5 inches long; the tube will break easily if a single scratch is first made with the edge of a three-cornered file. These tubes are to be used as open tubes, as explained later. The size with the smaller bore should be cut into 6-inch length and held with the middle point in the hot part of the Bunsen flame. When the glass is soft, draw the two ends apart by a quick motion (without twisting), and then heat each long tapering end in the flame and pinch it off short while it is hot, using the steel end of the forceps. In this way two closed tubes will be made from each piece; several should be made and kept in a closed box ready for use. A tube must be clean inside and out and should not be used twice.

7. *Fluxes and Other Chemical Reagents.* The chemical reagents needed are the fluxes* borax (sodium tetraborate), soda (sodium carbonate), and salt of phosphorus or microcosmic salt (phosphate of soda and ammonia). Each of these may be kept in a small bottle. Small bottles of hydrochloric, nitric, and sulfuric acids are also useful, and one of a solution of cobalt nitrate, as well; each bottle may conveniently have a glass dropping tube with a bulb in place of the ordinary glass stopper.

The following articles also will be found very convenient.

A small hammer with a square face with sharp edges and a steel anvil an inch or two long.

A horseshoe magnet (Fig. 185), the place of which may be taken by a magnetized knife blade.

A small agate mortar and pestle; also a steel diamond mortar (one in which the pestle fits tightly) in which a hard mineral can be pulverized without loss of the fragments.

A three-cornered file.

A few small watch-glasses, several small dishes of glass or porcelain to hold the fragments of the mineral under examination, and several test tubes are desirable. If chemical tests proper are to be tried, a wash bottle (for distilled water), a bottle of ammonia, and some filter paper.

* So called because they help in the melting or fusion of the substance under examination.

FIG. 185. Horse-shoe Magnet.

Before beginning to experiment cover the table with a piece of asbestos wallboard and place the lamp and reagent bottles upon it. The student must remember also that the acids mentioned are powerfully corrosive and will stain and finally destroy any fabric, such as clothes or carpet, that they are allowed to touch.* Moreover, the fumes from the acids when hot are injurious, and, therefore, for any extended series of strictly chemical tests it is almost essential to have some of the conveniences of a laboratory. Still another caution is needed; do not put away a piece of charcoal after use until it is quite certain that no fire lingers in it.

### HOW TO USE THE BLOWPIPE

The first thing to learn in the use of the blowpipe is to blow a hot, steady flame. Place the tip of the blowpipe close to or just within the flame as shown in Fig. 186, directing it slightly downward,

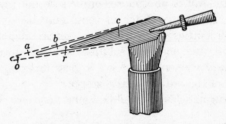

FIG. 186. Blowpipe Flame.

and blow through the tube. The blast of air will direct the flame into a thin cone, and with a little practice a clear blue flame, quite free from yellow, will be the result. This flame is much hotter than the ordinary gas flame, and, when the blowpipe is in skillful hands, it is hot enough to melt a fine platinum wire. The hottest part is at the extremity of the blue flame (at *a* in Fig. 186).

It may be difficult at first to blow a continuous, steady flame, but

* In case of accident the effect of the acid can often be neutralized by the prompt application of ammonia or carbonate of soda, which may afterward be washed out with a little water.

the ability will be acquired by practice. It is only necessary to continue to breathe slowly through the nose while the pressure of the cheeks upon the reservoir of air kept all the time in the mouth prolongs the blast. This pressure need not be great—not enough to tire the cheek muscles perceptibly except after a long time; if fatigue comes soon, it is because the student is unskillful or has a bad blowpipe.

An important distinction must be made between the *reducing flame* and the *oxidizing flame*. The flame in general consists of two parts: the inner blue cone, and the outer almost invisible envelope extending far beyond. In the inner cone the gas is only partly burned, and there is a deficiency of oxygen. (Point $b$, Fig. 186.) A substance placed here will part with its oxygen and be *reduced*, as when oxide of nickel, NiO, is changed to metallic nickel (Ni).

In the outer part of the flame, on the other hand, there is an excess of oxygen from the surrounding air (point $a$, Fig. 186), and the tendency is to give off oxygen or to oxidize a mineral placed within it. Here the lower oxide of manganese, MnO, for example, is changed to the higher oxide $Mn_2O_3$.

This distinction between the action of the two parts of the flame is very important in a certain class of experiments. The student must notice further that to blow a strong oxidizing flame the tip of the blowpipe should be placed just inside the gas flame, as indicated in Fig. 186; the flame is then free from yellow, and the substance under experiment should be held well beyond the end of the blue cone, at $o$.

For a reducing flame, on the other hand, the tip should be placed a little outside the gas flame, so that a little yellow follows the flame above the blue cone; the substance is held at $b$ within the blue cone and for best results is more or less surrounded by the yellow flame.

In succeeding pages the different methods of examination with the aid of the blowpipe are briefly described. The student should take them up in order, going through as many as possible of the trials with the minerals suggested and endeavoring to obtain the results described as closely as he can. It is essential that *small* fragments of pure material be used for the experiments, for, although the temperature of the blowpipe flame is high, the amount of heat is not great enough to produce the desired results with a large fragment.

## EXAMINATION IN THE FORCEPS

A small fragment of a mineral, held in the platinum points* of the forceps, may be tested to see whether it can be melted, and if so, whether easily or with difficulty. At the same time one should observe whether the mineral imparts a color to the flame, for the color will give information about the composition of the mineral.

As a first experiment, take a little sliver of barite; place it between the platinum points, letting the edge project well beyond them; blow a clean blue flame with the blowpipe, and just in front of this (in the oxidizing flame, see Fig. 186) insert the mineral. It will melt rather easily to a white opaque glass, and at the same time the flame beyond will be streaked with a pale yellowish green that is characteristic of the element barium.

If a piece of barite crystal is employed, it is very likely to break violently into fragments when it is introduced into the flame. This action, called *decrepitation*, is not uncommon, especially with crystallized minerals. It can often be prevented by heating the fragment slowly at first, but sometimes it is necessary to begin by powdering the mineral, mixing it with a drop of water, and finally supporting the thick paste so formed on a loop at the end of the platinum wire.

**Scale of Fusibility.** The method of experiment described gives in the first place an approximate determination of the melting point or degree of fusibility. The following scale is used to define the fusibility of the different minerals.

1. *Stibnite:* fusible in the ordinary yellow gas flame or candle flame even in large fragments.
2. *Natrolite:* fusible in fine needles in the ordinary gas flame, or in larger fragments in the blowpipe flame.
3. *Almandite*, or iron-alumina garnet: fusible to a globule without difficulty with the blowpipe, if it is in thin splinters.
4. *Actinolite:* fusible to a globule in thin splinters.
5. *Orthoclase:* thin edges can be rounded without great difficulty.
6. *Bronzite:* fusible with difficulty on the finest edges.

The accompanying list gives the names of some minerals, most of them common, with the degree of fusibility of each according to this scale.

* If platinum-tipped forceps are used, one should avoid holding arsenic and antimony minerals with them for these minerals make fusible alloys with the platinum. Forceps with Nichrome tips are more satisfactory for the beginner.

| | |
|---|---|
| Stibnite, galena | 1 |
| Cryolite, apophyllite, pyromorphite | 1½ |
| Amblygonite, witherite, prehnite, arsenopyrite | 2 |
| Rhodonite, analcime | 2½ |
| Gypsum, barite, celestite, fluorite, epidote | 3 |
| Oligoclase | 3½ |
| Albite | 4 |
| Apatite, hematite, magnetite | 5 |
| Bronzite | 6 |
| Quartz, calcite, topaz, sphalerite, graphite | Infusible |

The student must be warned that the method of expressing the fusibility of a mineral by referring it to the scale given is not exact. The results obtained in different cases will depend upon the size and the shape of the fragment employed, the conductivity of heat, and the skill of the experimenter.

**Flame Coloration.** Besides the fusibility, this experiment with a fragment of barite in the forceps serves to prove the presence of barium by the color given to the flame. It is found that a considerable number of substances are characterized in the same way; hence the flame coloration becomes a simple and important means of qualitative blowpipe chemical analysis. The following is a list of the flame colors produced by some of the common elements:

FLAME COLORATIONS

| *Color* | | *Element* |
|---|---|---|
| Red | Carmine-red | Lithium |
| | Purple-red | Strontium |
| | Yellowish red | Calcium |
| Yellow | | Sodium |
| Green | Yellowish green | Barium |
| | Siskine-green | Boron |
| | Emerald-green | Copper |
| | Bluish green | Phosphates |
| Blue | Greenish blue | Antimony |
| | Whitish blue | Arsenic |
| | Azure-blue | Copper chloride |
| Violet | | Potassium |

It may be noted here that the blue flame of copper chloride is sometimes used as a test for chlorine. For example, if powdered pyromorphite is mixed with a little powdered cuprite, and the mixture is fused together upon charcoal, a blue flame will be obtained for a moment, indicating that the pyromorphite contains some chlorine.

The flame color characteristic of a given element is often masked by another; thus the green of the boron in borax is concealed by the stronger yellow of the sodium, but it may be seen clearly if a drop of sulfuric acid is placed on the substance before heating. Further, a difficultly fusible or an infusible mineral is often not sufficiently decomposed by heat alone to show the flame color, and hence a more complex method is required; thus a fragment of apatite, first moistened by sulfuric acid, gives in the forceps the green flame of phosphorus. Again, to obtain the potassium flame from ortho-clase, a fragment of orthoclase should be mixed with an equal bulk of powdered gypsum and made into a paste with a little water. When the mixture is fused on a clean platinum wire, the violet flame of potassium becomes visible. If the violet is masked by a strong sodium flame, a piece of blue glass will absorb the yellow and allow the violet to be seen.

In testing the carbonates, such as calcite and strontianite, a drop of hydrochloric acid will result in the formation of the chloride (of calcium and strontium), which gives a more intense color to the flame.

At the same time that he tests the fusibility and flame color of a mineral, the student must be on the watch for attendant phe-nomena. For the fragment, instead of fusing quietly, may:

(*a*) *Swell up;* that is, throw out little globules or curling processes (as stilbite).

(*b*) *Intumesce;* that is, bubble up and then fuse (as scapolite and most zeolites).

(*c*) *Exfoliate;* that is, swell up and open out in leaves (as apophyllite and, even more, vermiculite).

(*d*) *Glow brightly*, without melting (as calcite).

The fragment after being heated should be examined to see whether, if fused, the glass is clear, full of bubbles (then often called *blebby*, or *vesicular*), or even black; whether it has changed color, even if not fused; whether it is magnetic (due to iron).

Finally, a few infusible substances turn blue when, after being heated, they are moistened with a drop of cobalt nitrate solution and again heated. This is true of kyanite, kaolin, and other infusible minerals containing alumina. Hemimorphite (zinc silicate) also turns blue under similar circumstances. A blue obtained when fusible minerals are treated in this way may be due simply to the cobalt (see p. 110).

## USE OF THE PLATINUM WIRE

Wind the platinum wire around a small card* leaving about 3 inches free; then bend the end into a loop about $\frac{1}{8}$ inch across (Fig. 187); it is now ready for use. Heat the loop in the blowpipe flame and dip it in the borax; some of the borax will adhere, which may then be fused to a colorless glass. Add a little more, and repeat the operation until the bead is clear and round and fills the loop entirely but not until it is so big that it will fall off the wire.

Fig. 187. Loop in Platinum Wire.

Now take a few fine fragments of cuprite or malachite, heat the bead, and bring it into contact with one of them; it should adhere, and, when it is heated again in the oxidizing flame, it will slowly dissolve and disappear (note the emerald-green flame of the copper). If the quantity of the copper mineral was very small, the bead will be found to be clear green while hot, and blue when cold. If a larger amount was taken or is now added, and the bead is again heated, it may remain green on cooling, or even be so deeply colored as to appear black and opaque. In the last case the color may often be seen if the bead is heated hot and quickly flattened out by the pressure of a hammer. It is always best to commence with a minute portion and then add more, rather than to commence with so much that the bead is black. If too much has been taken, it is best to shake off the bead from the wire by a sudden motion when it is hot, or break it when it is cold, and begin again.

Suppose now that the bead is deep green and contains a relatively large amount of copper or is saturated. Hold it in the reducing flame (see Fig. 186) and heat again. Now the oxygen needed to burn the gas will be taken in part from the oxide of copper ($CuO$) in the bead, and part of it will be changed or reduced

* Instead of this, the wire may perhaps better be cut into short pieces about 3 inches or less in length and each fused into a glass tube drawn out as described on p. 103.

to the lower oxide of copper, $Cu_2O$, which will show red and opaque when the bead is cold. The color of the borax bead thus is a test which can prove the presence of copper, and similarly also the presence of a number of the other metals can be proved.

The following is a list of the common metallic oxides and the colored beads that they yield. The distinction between the oxidizing flame (O.F.), which is to be used first, and the reducing flame (R.F.) is to be carefully observed. Thus nickel gives a violet bead in the O.F. from the oxide NiO, whereas in the R.F. it becomes gray and muddy from metallic nickel.

It may be added that many of the common metals, such as silver, zinc, and lead, give colorless beads in both the O.F. and the R.F.

### BORAX BEAD COLORS

| Hot O.F. Color | Cold O.F. Color | Hot R.F. Color | Cold R.F. Color | Element |
|---|---|---|---|---|
| Pale yellow | Colorless to white | Grayish | Brownish violet | Titanium |
| Pale yellow | Colorless to white | Yellow | Pale yellow | Tungsten |
| Pale yellow | Colorless to white | Brown | Brown | Molybdenum |
| Yellow | Yellowish green | Green | Green | Chromium |
| Yellow | Yellow-green to colorless | Dirty green | Green | Vanadium |
| Yellow to orange-red | Yellow | Pale green | Pale green to colorless | Uranium |
| Yellow to orange-red | Yellow | Bottle green | Pale bottle green | Iron |
| Pale green | Blue-green | Colorless to green | Opaque red | Copper |
| Blue | Blue | Blue | Blue | Cobalt |
| Violet | Red-brown | Opaque gray | Opaque gray | Nickel |
| Violet | Reddish violet | Colorless | Colorless | Manganese |

Before introducing a mineral fragment into the borax bead the process of roasting should be performed to expel the arsenic, antimony, or sulfur. Roasting consists in heating the powdered substance on charcoal cautiously, so as not to fuse it, first in the O.F., then in the R.F., and repeating this a number of times patiently. The mineral should still be in the state of a powder at the end.

It must be noticed that, when two metals are present, the color of one may not be seen; thus, if iron, nickel, and cobalt are in the substance under examination, the colors are observed in the order given, if each metal in turn is oxidized by skillful treatment on charcoal with successive portions of borax.

Soda is particularly useful on charcoal in reducing the metallic compounds as described later. With manganese in the oxidizing flame soda gives a fine bluish green but opaque bead. This is the best test for manganese. The soda beads in general are opaque, but with silica a clear glass may be obtained.

FIG. 188. Antimony Sublimate on Charcoal.

## USE OF CHARCOAL

A piece of charcoal such as has been described on p. 102 is very useful in the chemical examination of minerals with the blowpipe. It forms a support for the mineral being heated and a place upon which a deposit of a volatile compound may be formed. Besides this the glowing carbon has what has been called (p. 105) a powerful reducing effect; that is, it takes oxygen (or other elements) away, and sometimes without other means this suffices to produce the metal from its compound. Thus cuprite ($Cu_2O$) is reduced to metallic copper by heating on charcoal; similarly chalcocite ($Cu_2S$) may be treated with the same result, and cerargyrite (AgCl) yields metallic silver.

For certain tests it is more desirable to use a plate of plaster of Paris as a support in place of the charcoal.

The way in which the charcoal is used and the chemical principles involved will be made clear by a few examples.

The fragment of mineral used should be placed near one end of the rectangular piece of charcoal (see p. 102 and Fig. 188) and held so that the flame will sweep down the full length. If a volatile substance is formed during the heating, it will have the best opportunity to deposit on the cooler surface. It is not necessary to make a deep hole; often a short scratch made across the charcoal with a sharp edge is sufficient; the fragment is blown against this by the flame. When the fragment persists in jumping off, it may sometimes be held in place by fusing a little borax to it.

As an experiment, take a piece of stibnite (sulfide of antimony, $Sb_2S_3$), place it on the charcoal, and heat gently. It will fuse very easily and will give off a cloud of white fumes. These collect as a white coating over the coal; the black surface seen through the white coating will give the effect of blue on the edges. If the heating is continued, the mineral will entirely disappear, or, in other words, it is entirely volatilized. Such a coating is called a sublimate, and, in this case, it consists of the antimony trioxide ($Sb_2O_3$) formed by the union of the antimony with the oxygen of the air. At the same time the sulfur combines with oxygen and goes off as a gas ($SO_2$). If now the reducing flame is thrown for a moment against the white coating, the coating is burned off with a bluish flame. The action of the flame is to reduce the oxide to the metal (Sb), which is instantly volatilized; as it goes off, it is again oxidized.

For a second experiment take a fragment of orpiment (sulfide of arsenic $As_2S_3$) and treat it in the same way. The result is somewhat similar; it fuses easily, giving white fumes of $As_2O_3$; it is also entirely volatile. But now a strong disagreeable odor will be perceived as the fumes are formed; this is usually described as a garlic odor; it is characteristic of arsenic and is always produced when the metal is volatilized and arsenic trioxide ($As_2O_3$) is formed. The odor serves to distinguish arsenic from antimony; moreover, the white coating will be seen this time to lie much farther from the flame than the oxide of antimony because it is more volatile and can be deposited only where the coal is comparatively cool.

A third trial may be made with arsenopyrite. It gives off, as it is heated, a cloud of white fumes with the same peculiar penetrating, garlic odor, and the white coating of arsenic trioxide forms at a distance on the coal. There is, however, a residue in this case which soon fuses to a grayish black globule that, when cold, is found to be magnetic, thus proving the presence of iron. The mineral consists of iron, sulfur, and arsenic (the formula is FeAsS). Part of the sulfur is driven off (as $SO_2$), and, after some time, all the arsenic also is driven off, while a magnetic compound of iron and sulfur (with perhaps a little residue of arsenic) is left behind.

A fragment of galena should also be tried on charcoal. It will fuse very easily, and immediately a yellow coating of the oxide of lead (PbO) will form about it, while farther away a white coating of lead sulfate ($PbSO_4$) will be formed by the union of the PbO and the $SO_2$ in the presence of the oxygen of the air.

Further, lead is what is called an easily reducible metal; that is, its compounds are rather easily changed to the metallic state under the action of heat. Hence, continued heating yields a globule of metallic lead. A little soda on the galena hastens the production of the metal, and at the same time it is noted that the yellow coating is more distinct, whereas the white fumes are nearly absent, for now the soda unites with the sulfur of the galena.

A mineral containing copper will yield metallic copper on charcoal when it is heated with the soda. This may be either in small globules or as a thin crust. When exposed to the air the copper becomes coated with the black oxide but is easily recognized since it is malleable on the anvil and shows its peculiar red color when rubbed.

The examples outlined above and some reactions of other elements on charcoal are summarized in the following table.

## REACTIONS ON CHARCOAL

*Arsenic* gives a white, volatile deposit some distance from the assay. (See description of orpiment and arsenopyrite p. 112.)

*Antimony* gives a white, volatile deposit close to the assay. (See description of stibnite, p. 112.)

*Selenium* gives a white, volatile deposit tinged with red on the outside, accompanied by a disagreeable odor. If the coating is touched with the reducing flame, it gives a blue flame.

*Tellurium* gives a dense white, volatile coating which, when touched with the reducing flame, gives a bluish green flame color.

*Zinc* minerals should be powdered and mixed with two or three times their volume of soda before heating on charcoal. The sublimate is nonvolatile, yellow when hot and white when cold. If the coating is moistened with cobalt nitrate and again heated, it turns green.

*Tin* forms very few natural compounds. The only tin mineral of importance is cassiterite, and it is extremely infusible and refractory. It should therefore be powdered and mixed with soda before heating on charcoal. After prolonged heating there will be minute malleable globules of *metallic tin*. These are at first nearly as white as silver, but they soon oxidize and become dull. With a little nitric acid in a watch glass they yield an insoluble white powder of tin dioxide.

*Molybdenum* gives a sublimate that is pale yellow when hot and white when cold, and volatile in the oxidizing flame. If the sublimate is touched with the reducing flame, it turns a beautiful azure-blue. This blue is difficult to see on the black charcoal, and thus it may be desirable to make this test on a plaster block.

*Lead* gives a yellow coating near the assay and white farther away. (See description of galena test p. 112.) Most lead minerals are easily reduced and give a globule of metallic lead.

*Bismuth* gives a yellow sublimate near the assay and white farther away, much resembling lead. If the bismuth mineral is mixed with a little potassium iodide and sulfur (bismith flux) and again heated, a bright red coating results. A lead mineral under the same conditions gives a deep yellow coating.

## USE OF THE CLOSED AND OPEN TUBES

The tubes in blowpipe work are used chiefly in the examination of minerals that yield on heating a volatile substance which in most cases will be condensed on the colder part of the tube. An important distinction is to be made between the use of the closed and the open tube. The closed tube contains very little air, and this is driven out with the first puffs of gas from the heated mineral, and hence any reaction takes place with little or no effect from the oxygen of the air. In the open tube, if it is held in the proper inclined position, a constant stream of hot air (that is, of oxygen) passes up the tube and over the heated mineral fragment. A few examples will show how this principle is applied.

Place a little fragment of sulfur in the closed tube and heat it gently. At once it is fused and converted into sulfur vapor which rises in the tube and soon condenses, giving a dark orange-red ring of liquid sulfur, which becomes light yellow as it cools and solidifies. Here there has been no change, simply the volatilization of the sulfur.

Now place a fragment in an open tube, about an inch from the end; incline the tube as much as possible without causing the fragment to slip out, and heat it very slowly. The sulfur fuses as before, but the hot oxygen which passes over it unites with it, forming sulfur dioxide ($SO_2$), an invisible gas which rises through the tube and comes out of the open end, giving the usual sulfur odor. Further, the acid fumes of this gas will turn a piece of blue litmus paper bright red. It is difficult to heat the sulfur slowly enough to prevent the formation of a ring also, as in the closed tube, simply because sulfur is easily volatile; that is, it goes off into gas very readily, and the oxygen cannot be supplied fast enough to oxidize all of it.

As a second example, take a small piece of as pure cinnabar as can be obtained. The cinnabar is sulfide of mercury, HgS, a substance that is converted into vapor when heated out of contact with the air.  In the closed tube we get at once a black ring, or sublimate, of mercury sulfide which, like the sulfur, was first volatilized and then condensed where the tube was cooler.  This black coating becomes reddish if rubbed.

Now place a fragment of cinnabar in the open tube and heat it very slowly and carefully.  Gradually the cinnabar disappears, while the sulfurous fumes can be perceived at the end of the tube, as in the case of the pure sulfur.  In addition, a little above the fragment a faint deposit begins to form, growing more and more distinct, and, finally, when seen by reflected light, it appears as a shining mirror.  This is metallic mercury in the form of minute globules coating the glass; that it is mercury can be proved even to the skeptical by rubbing the deposit with a matchstick.  The minute globules unite to form a few large ones which will run out of the tube.

It is not difficult to explain what has happened in this experiment.  The hot oxygen passing over the heated mineral has united with the sulfur to form sulfur dioxide ($SO_2$), while the mercury thus left alone has been driven off as vapor by the heat and collected where the tube was cool enough to allow its condensation.  Very likely in this instance too, unless the heating is very slow, a little sulfide of mercury will go off without oxidization and form a black ring in the open tube, but, by gradually heating this, keeping the tube in the same position, it is driven up the tube while more and more of the sulfur is oxidized.  Finally nothing but the pure metallic mirror of the mercury is left.

This experiment succeeds best if the tube is first heated a little above the mineral and then the heating of the fragment is carried on very slowly and carefully.  If the powdered cinnabar is mixed with soda (first dried to expel the water), and then introduced into the closed tube and heated, a sublimate of metallic mercury is very readily obtained.

A fragment of galena placed in the closed tube undergoes no change and no sublimate is formed.  In the open tube, however, although no sublimate is produced, some of the sulfur is oxidized, and the sulfurous fumes can be perceived by the odor or by their reddening effect on litmus paper.  This method is consequently a general method of testing for sulfur in the class of compounds called sulfides.

A fragment of orpiment, sulfide of arsenic, $As_2S_3$, heated in the closed tube is melted, volatilized, and forms a beautiful red ring of sulfide of arsenic. Heated in the open tube (very slowly), both sulfur and arsenic are oxidized; the sulfur gives as always $SO_2$, while the arsenic yields a white deposit of minute octahedral crystals of arsenic trioxide ($As_2O_3$) spangling in the light. This sublimate is very volatile and hence may be driven farther and farther up the tube when heated.

A mineral containing water, when heated in a closed tube, gives off the water vapor that condenses as droplets in the upper part of the tube. A change in the appearance of the mineral may take place at the same time. Thus a piece of goethite, or hydrated oxide of iron, gives off its water and turns red, for it is now the anhydrous oxide of iron, like hematite, which has a red powder. In a few cases, such as some sulfates, the water has an acid reaction and turns blue litmus paper red.

The more important tube tests are briefly summarized below.

### OPEN AND CLOSED TUBE TESTS

**Sulfur.** When native sulfur or a sulfide containing a high percentage of sulfur is heated in a closed tube, a sublimate forms that is red when hot and yellow when cold (see p. 114). In an open tube sulfur dioxide, a colorless gas, issues from the upper end and can be noted by its characteristic odor. A moistened piece of blue litmus paper placed in the upper end turns red. This is a general test for all unoxidized sulfur compounds (sulfides and sulfosalts).

**Arsenic.** The reaction of realgar in a closed tube is given on p. 115; the same results are obtained with the other arsenic sulfide, orpiment. Other arsenic compounds give two rings around a closed tube. One is black; the other near the bottom is silver-gray crystalline arsenic, the "arsenic minor." In an open tube a white, highly volatile sublimate of arsenic oxide forms on the cold part of the tube. The characteristic garlic odor will also be noted.

**Antimony.** In a closed tube antimony compounds give a slight reddish brown coating near the bottom of the tube. In an open tube sublimates of two antimony oxides usually form. One is a white volatile ring close to the heated position; the other, which is pale yellow and non-volatile, settles mostly on the bottom of the tube

**Mercury.** In a closed tube mercury (cinnabar) gives a black sublimate of mercury sulfide, but, mixed with sodium carbonate and heated, gray metallic globules of mercury condense on the cool walls of the tube. In an open tube, if it is heated slowly, the sulfur is oxidized and globules of metallic mercury remain. (See p. 115.)

**Water.** Any mineral containing water will give, on heating in the closed tube, a deposit of drops of water on the cold walls of the tube. If the mineral contains hydrochloric, sulfuric, or other volatile acid, the water will turn blue litmus paper red. As an experiment, try gypsum first and then alunite. The water from the gypsum is neutral, but that from the alunite will give a strong acid reaction.

## CHEMICAL EXAMINATION BY ACIDS AND OTHER REAGENTS

In addition to the various methods of chemical examination already described that can be made by means of the blowpipe, there are a few other simple chemical tests that may occasionally be necessary. The reagents most needed are the three acids, hydrochloric, nitric, and sulfuric, and a little ammonia. Usually undiluted acids are preferable, but often acids diluted with an equal volume of water are adequate. A few test tubes, preferably of hard glass, are also required. The caution already mentioned (p. 104) in regard to chemical reagents should be carefully observed.

**Solubility in Acid.** Whether a mineral is soluble in one of the acids named is often very important. To test the solubility hydrochloric acid is generally used, except with metallic sulfides and some other minerals containing predominantly one of the heavy metals (lead, copper, silver, etc.); for these, nitric acid is usually better. In general the mineral should be pulverized as finely as possible in the agate mortar and introduced into a large test tube, some acid poured on, and the whole carefully heated over the Bunsen flame.

It must be remembered here that it is injurious to breathe the acid fumes in the air and that they will act corrosively upon surfaces of brass near by; hence such tests can be tried only with caution unless the conveniences of the laboratory are at hand.

Various results may be noted during this trial:

A. The mineral may dissolve quietly with or without coloring the solution; this is true, for example, of hematite, also of many of the sulfates and phosphates.

B. There may be a bubbling off of effervescent gas. This gas is usually carbon dioxide ($CO_2$); but it may be hydrogen sulfide ($H_2S$).

C. There may be a separation of some insoluble substance, such as sulfur or silica. These points will now be discussed more in detail.

**Effervescence Yielding Carbon Dioxide.** This type of effervescence resembles the bubbling observed in a glass of soda water and is due to the escape of the same gas. It is an easy and important test for the carbonates. Some carbonates dissolve in cold acid even if they are in lumps that have not first been pulverized. Calcite, aragonite, strontianite, withenite, and smithsonite belong to this group of carbonates, but dolomite and siderite, which must be pulverized or heated, or both, do not; hence, this test is a means of distinguishing between them. The carbonates of copper and lead should be tried with nitric acid.

**Effervescence Yielding Hydrogen Sulfide.** Most metallic minerals, as stated above, should be treated with nitric acid, but some that do not have a metallic luster, sphalerite, for example, may be put into hydrochloric acid. With sphalerite the reaction produces the gas hydrogen sulfide, while zinc chloride goes into solution. The gas bubbles off like carbon dioxide, but its disagreeable odor, resembling that of rotten eggs, shows at once what it is.

**Chlorine,** easily detected by its peculiar odor, is sometimes given off, as when the oxides of manganese are heated in hydrochloric acid.

**Separation of Sulfur.** A number of sulfides, such as pyrite, dissolve in nitric acid with the separation of particles of sulfur that usually cling together and float on the liquid. Chalcopyrite reacts in this way, but, like other copper sulfides, it gives a green solution which turns a deep, fine prussian blue when ammonia is added in sufficient quantity to dissolve the precipitate that forms at first.

**Separation of Tin Dioxide.** When metallic tin is treated with nitric acid, tin dioxide ($SnO$) is formed, which separates as an insoluble white powder.

**Separation of Silica.** A number of silicates dissolve in hydrochloric acid with the separation of the silica, sometimes as a powder, sometimes as a slimy mass. Other silicates dissolve entirely, but, if the solution is gently heated until part of the liquid has been evaporated, a thick jelly is finally formed, so that the test tube can be partially inverted without its flowing out. Such silicates are said to gelatinize with acid. This is true of hemimorphite and a number of the zeolites; chabazite, on the other hand, is decomposed with the separation of the slimy silica.

**Difficultly Soluble or Insoluble Minerals.** A large number of minerals, even when pulverized, dissolve very little or not at all in strong hot acid. Quartz, corundum, orthoclase, topaz, and many others, even when they are finely pulverized and long heated in strong acid, are not at all, or only very slightly, attacked. The question whether there has been partial solution is not always easy to answer, but it can be decided if the liquid takes a distinct color, or it can be more fully decided if the liquid is filtered off from the undecomposed mineral, and then a few drops of ammonia are added to it. This procedure will generally cause the bases which have gone into solution to separate as precipitates. To explain the various ways—though many of them are simple— in which the bases present in the solution can be identified would take us too far into the subject of chemistry. Do not forget, however, the test for copper just mentioned (p. 118). Attention may be called to the further fact that, as a test for sulfuric acid or a sulfate, the addition, to a solution containing them, of a little barium chloride will cause a heavy white precipitate of barium sulfate to form.

# CHAPTER VII

# Description
# of the Mineral Species

In this chapter are given descriptions of all the common minerals with brief remarks about some of those which are rarer. The classification outlined on p. 122 is the same as that used in more advanced works on mineralogy in which the minerals are arranged in chemical groups.

In general the following properties are considered for each mineral:

**Habit.** Under this heading are given the crystal system, characteristic forms, type of crystalline aggregate, and massive varieties.

**Physical Properties.** Included here are cleavage, and, if important, fracture and tenacity, hardness, specific gravity, luster, color, and streak. Other physical properties that are important for some minerals, such as fluorescence and magnetism, are also given.

**Composition.** Under this heading are given the chemical composition and any properties that are directly dependent on chemical variations. Blowpipe and chemical tests that are useful in identifying the mineral are also given.

**Occurrence.** The type of geologic occurrence is given for those species for which it is characteristic. Space does not permit a complete listing of localities where the mineral is found, but a few of the more important ones are given.

Preceding the description of most of the minerals is a brief statement of some of the more important uses and any striking or unusual features. Attention is directed to facts about the mineral that are most easily remembered so that the student will have before him facts with which to associate the other properties of the mineral.

In the description of many species no mention is made of certain properties which are relatively unimportant in these particular cases. Thus, if the cleavage is not mentioned, it is because it either is not observed or is too imperfect to be important. Since nearly all minerals are brittle, it is unnecessary to repeat this word in each mineral description; but, if the mineral is not brittle but malleable or sectile, this property is stated and should be carefully noted. If the streak is not given, it is understood to be *white* or nearly white, like that of most nonmetallic minerals, even when the mineral itself has a deep color. All minerals having a metallic luster are opaque.

The student will find it easier to remember the properties of different minerals if, after studying the descriptions in the book and comparing them with specimens to which he has access, he makes a tabular list of the properties for each species, somewhat similar to the suggested form.

SUGGESTED FORM FOR TABULATING THE PROPERTIES OF MINERALS

| | DIAMOND | GRAPHITE | GALENA | SPHALERITE |
|---|---|---|---|---|
| Crystal system and common form | Isometric octahedron | Hexagonal tabular | Isometric cube | Isometric tetrahedron |
| Aggregates | ......... | Foliated | Granular-cleavable | Granular-cleavable |
| Cleavage | Octahedral | Basal | Cubic | Dodecahedral |
| Hardness | 10! | 1–2! (flexible) | 2½–3 | 3½–4 |
| Specific gravity | 3.5 | 2.2 | 7.5! | 4 |
| Luster | Adamantine | Metallic | Metallic | Resinous |
| Color | Colorless, yellow | Black | Lead gray | Yellow, brown, black, etc. |
| Streak | White | Black | Dark gray | White to brown |
| Composition | Carbon | Carbon | PbS | ZnS |
| Tests | Infusible | Infusible | Easily fusible | Infusible |

It is easy to arrange a notebook for this purpose by ruling a series of parallel vertical columns, and, to avoid writing the list of properties on each page, they may be written on the edge of the first left-hand page and the corresponding strip from a sufficient number

of the subsequent sheets may be neatly cut off. When a property
is particularly important it should be underscored or followed by
an exclamation point. It is not worth while to repeat in tabular
form the entire description in the text; a little experience will soon
show how much may be advantageously written down.

Filling out a similar column of properties determined from the
specimen itself and comparing it with the list in the student's
notebook made out from the text is a useful exercise. If the species
was unknown at first, the list of properties will often suffice to
determine it.

It is not necessary to *learn* by sheer effort of memory all the
properties at once; this would be difficult and tiresome. The
important properties can be learned (such as chemical composition), but the knowledge of most of the physical properties will
be acquired gradually by repeated handling of the specimens
themselves.

## CLASSIFICATION

Several methods have been used to classify minerals, and each
has its value depending on the properties one wishes to emphasize.
An arrangement in common use in books on beginning mineralogy
is according to the prominent elements of which the minerals are
compounds. In this way all the iron minerals, such as hematite,
goethite, magnetite, and siderite, are grouped together. Likewise
the zinc minerals—sphalerite, smithsonite, hemimorphite, and
willemite—are together. This may seem to the beginner the best
and most logical way to classify minerals, and indeed it might be
if we were considering minerals only from an economic viewpoint
as the source of the metals they contain. For those who wish to
consider them in this manner, a list of the minerals arranged according to the most important element is given in Appendix I,
pp. 309 to 311.

To the trained mineralogist the presence of a certain metal
is not nearly so important as the type of compound and the properties of the compound. For example, the native elements copper,
silver, and gold, besides being metals, are isometric with similar
habits and physical properties. The group of carbonates known
as the *calcite group* is an even better example:

Calcite, $CaCO_3$                     Siderite, $FeCO_3$
Dolomite, $CaMg(CO_3)_2$              Rhodochrosite, $MnCO_3$
Magnesite, $MgCO_3$                   Smithsonite, $ZnCO_3$

All these minerals have the same type of crystal structure, and the angles between the corresponding crystal faces are nearly the same, varying from 105° to 107° over the edges of the rhombohedron. Hence they are called *isostructural*, a term that means that they have similar crystal structure and analogous chemical composition. If they were classified according to the metal present in each, they would be scattered through the book, and these similarities would not be brought out. Calcite would be found with apatite, calcium phosphate; with fluorite, calcium fluoride; with anhydrite, calcium sulfate; and with others with which calcite has nothing in common except the presence of calcium. All the other carbonates would be scattered also. Furthermore, it is difficult to place some minerals; for example, dolomite could be classified equally well with either the calcium or the magnesium minerals.

For these reasons the minerals described later in this book are grouped according to a chemical classification in which all the sulfides, all the oxides, all the sulfates, etc., are grouped together. This is the classification used by all modern advanced textbooks on mineralogy. Furthermore, it seems wise for the beginner to start with a classification with which he must be familiar when he undertakes advanced work.

The chemical classes into which the minerals are grouped are as follows:

1. *Native Elements.* A few of the elements occur in nature uncombined and are hence called native elements, such as native gold and native sulfur.

2. *Sulfides.* The sulfides are compounds of a metal with sulfur, such as galena, PbS; sphalerite, ZnS; pyrite, $FeS_2$. Similar to the sulfides and included with them are the rare tellurides and arsenides, such as calaverite, $AuTe_2$, and niccolite, NiAs. Also included with the sulfides are those minerals called sulfosalts. They are compounds composed of a metal in combination with sulfur and arsenic, antimony, or bismuth; for example, enargite, $Cu_3AsS_4$; and proustite, $Ag_3AsS_3$.

3. *Oxides.* The oxides are composed of a metal in combination with oxygen, such as hematite, $Fe_2O_3$; cuprite, $Cu_2O$; cassiterite, $SnO_2$. The hydrous oxides, in addition to the metal and oxygen, contain water or the hydroxyl (OH), as, for example, goethite, $FeO(OH)$.

4. *Halides.* The halides include compounds with chlorine.

fluorine, rare bromides, and iodides. Halite, NaCl, and fluorite, $CaF_2$, are examples.

5. *Carbonates*. The carbonates can be considered salts of carbonic acid, $H_2CO_3$, in which a metal takes the place of the two hydrogen atoms, such as calcite, $CaCO_3$; smithsonite, $ZnCO_3$; and cerussite, $PbCO_3$.

6. *Phosphates*. The phosphates are minerals that include the $PO_4$ radical in their fomulas. Grouped with them are the rare arsenates and vanadates. Amblygonite, $LiAlFPO_4$, and pyromorphite, $Pb_5Cl(PO_4)$, are examples.

7. *Sulfates*. The sulfates can be thought of as salts of sulfuric acid, $H_2SO_4$, in which a metal takes the place of the hydrogen. Also grouped with them are the tungstates, molybdates, and uranates. Examples are: barite, $BaSO_4$; anglesite, $PbSO_4$; and scheelite, $CaWO_4$.

8. *Silicates*. The silicates form a very large and complex chemical class of minerals in which one or more metals, perhaps four or five, are in combination with silicon and oxygen. Examples are orthoclase, $KAlSi_3O_8$, and rhodonite, $MnSiO_3$.

## NATIVE ELEMENTS

Of the ninety-two elements only a few are found in the native state, that is, uncombined with other elements. These are divided into three groups: (1) metals—the most important of which are gold, silver, copper, platinum, and iron; (2) semi-metals—arsenic, antimony, and bismuth; (3) nonmetals—of which carbon in the form of diamond and graphite and sulfur are the most important.

### GOLD, Au

Since the beginning of historic time, gold has been the most highly prized of all the metals. It is only within relatively recent years that other metals such as platinum and radium have had a value equal to or greater than that of gold. It was used by early peoples because it occurred in the native state and thus did not require an elaborate metallurgical process to extract it from the ore. Furthermore, it could be easily worked and fashioned into durable ornaments of pleasing color. Gold was used very early as a medium of exchange and has been used as such by all civilized people. It is interesting to note that, although there are other gold minerals, most of the gold of the world has been obtained from the native metal.

**Habit.** Gold is isometric, and at a few localities has been found in well-formed octahedral or dodecahedral crystals (Fig. 189). However, crystalline gold is usually in plates or in wirelike forms (Fig. 190). The plates are flattened parallel to an octahedron face and may show triangular markings, whereas the wires are usually elongated parallel to a three-fold symmetry axis. Crystals of any sort are the exception, and most gold is found in irregular masses which if large are called nuggets.

FIG. 189. Malformed Gold Octahedron.

**Physical Properties.** The physical properties of gold are so characteristic that they readily serve to identify it. The hardness is $2\frac{1}{2}$–3; it can be easily scratched by a knife, leaving a shining groove. The specific gravity of pure gold is 19.3. It is one of the heaviest substances and has a far greater specific gravity than any of the "fool's golds." However, this property is of little help in identifying gold in the average specimen, for the tiny flakes that

FIG. 190. Gold Crystals, Trinity County, California.

are present are greatly overshadowed by the large bulk of quartz or other minerals with which it is associated. Native gold may contain some silver, which lowers the specific gravity and makes the deep yellow of pure gold somewhat lighter. It is highly malleable and ductile and can be hammered out into sheets so thin that they will transmit a faint greenish light. As gold is the only yellow mineral that is malleable, it can be distinguished from all others. Chalcopyrite has a yellow color, but, if struck a blow with a hammer, it is crushed to a powder and not flattened like gold. Furthermore, if chalcopyrite is scratched by a knife, it yields a green powder, not the smooth groove of gold.

**Composition.** Most native gold contains some silver and may contain small amounts of copper and iron. Silver up to 16 per cent may be present; that from California carries between 10 and 15 per cent. Gold has long been known as a *noble metal*, since it is not attacked by ordinary acids. It is dissolved in *aqua regia*, a mixture of nitric and hydrochloric acids.

**Occurrence.** Gold occurs in all kinds of rocks of all ages on all continents. There are thus many types of gold deposits, but most commonly it is in veins associated with quartz. This gold-bearing quartz is often milky and may show little particles of gold scattered through it, but more often no gold is visible until it is crushed to a powder and washed to remove the lighter material.

In some places nature has been performing this crushing and washing process for many centuries, for, when gold-bearing rocks are disintegrated by weathering, the broken material is washed into the neighboring streams. The small particles of the lighter minerals are carried away, while the heavier gold works its way toward the bottom of the stream bed, forming a placer deposit. Much of the world's gold, particularly in the early days of gold mining, has come from the sands and gravels of stream beds, when the miner washed the gravels in his pan. Some placer deposits have been worked on a large scale by throwing a powerful stream of water against the gravel bank, a procedure that carries away the lighter rock and leaves the heavy gold particles behind. At the present time, most placer deposits are worked by giant dredges through which pass thousands of cubic yards of gravel a day and from which the gold is washed mechanically.

The greatest gold-producing country today is South Africa because of the large deposits of the Whitwatersrand, "The Rand," in the Transvaal. The mines extend to a depth of nearly 10,000 feet, the deepest in the world. In the United States gold is mined in many states, the principal producers being California and South Dakota.

## SILVER, Ag

Silver is one of the precious metals, equally useful for ornaments of many kinds, for utensils, and for money. The silver coins of the United States contain 9 parts of silver and 1 part of copper. One of the largest uses of silver is in light-sensitive compounds used in photography. Native silver is not uncommon, although the world's supply of the metal comes chiefly from other minerals.

FIG. 191.   Channarcillo, Chile.     FIG. 192.   Kongsberg, Norway.

Native Silver.

**Habit.** Silver is like gold in its habit. It is found in some places, though rarely, in distinct isometric crystals, and more frequently in arborescent or branching groups, in plates and scales or wirelike forms (Figs. 191 and 192). Most native silver, however, shows no crystal forms and is in irregular masses conforming to the shape of the cracks in the rocks in which it is found.

**Physical Properties.** Silver is highly malleable and ductile. The hardness is 2½–3. The specific gravity of pure silver is 10.5, but it is higher when silver is alloyed with gold as it often is. Silver is the best known conductor of both heat and electricity. When it is fresh, the color is tin-white with its surface having the appearance of a well-polished silver spoon. However, on exposure to the air, a tarnish forms which may be bronze to dull black.

**Composition.** Silver may contain some copper, gold, and mercury; more rarely traces of other metals. It is readily dissolved by nitric acid, forming silver nitrate. If hydrochloric acid is added to this solution, a white, curdy precipitate of silver chloride results. This reaction is a very delicate test for silver.

**Occurrence.** Native silver occurs in many places, but in only a few has it been found abundantly. The mines at Kongsberg, Norway, which have been worked for several hundred years, have yielded outstanding specimens of crystallized wire silver. In the early part of the twentieth century Cobalt, Ontario, was a large producer of native silver where it occurred associated with cobalt and nickel minerals. More recently it has been mined at Great Bear Lake, Northwest Territories, where it is associated with uraninite. In the United States native silver has been found with native copper in the Lake Superior copper mines. Montana, Colorado, and Arizona have also been producers.

## COPPER, Cu

Copper is one of the most useful metals and has been employed since early times in many ways both as a metal and in alloys. Today its greatest use is in electrical equipment because it is an excellent electrical conductor. Nearly all the power lines carrying electricity from the places where it is generated to our homes are copper; the windings on motors and other familiar electrical appliances are also copper. Copper is used in many alloys of which brass (copper and zinc) and bronze (copper and tin) are the best known. Much of the magnesium metal that is now coming into wide use is alloyed with small amounts of copper to give it more desirable characteristics. There are many other uses for copper, so many in fact that, next to iron, copper is used more extensively than any other metal.

**Habit.** Native copper is isometric, and octahedral and dodecahedral crystals have been found, but usually crystallized copper is in distorted arborescent groups (Fig. 193). It is usually found in irregular masses which show no crystal forms.

**Physical Properties.** Copper, like gold and silver, is highly ductile and malleable. The hardness is $2\frac{1}{2}$-3, and the specific gravity is 8.9. The reddish hue so characteristic of copper is called copper-red. However, this color is seen only on fresh surfaces, and, as the bright new cent soon becomes dull, so does native copper on exposure to the air.

**Composition.** Native copper may contain small amounts of silver, bismuth, mercury, arsenic, and antimony. The copper from Lake Superior contained extremely small percentages of these impurities, and thus *Lake copper* for many years was considered the standard for high purity. Copper is easily dissolved by nitric

Fig. 193. Native Copper, Lake Superior, Michigan.

acid, giving a blue solution, and an excess of ammonia added to this solution turns it a deep azure blue.

**Occurrence.** The most celebrated locality for native copper is on the Keweenaw Peninsula in Michigan on the shores of Lake Superior, where it has been mined since the middle of the nineteenth century. Today mining activity has almost ceased there, but in the past production was large. Beautiful crystallized specimens were found there associated with calcite, datolite, and a number of zeolites. In some specimens the copper is enclosed in crystals of calcite which imparts the bright red color to the crystals. Great masses of native copper have been found there, one of which weighed 420 tons.

Native copper is widely distributed in small amounts in the oxidized zones overlying deposits of copper sulfides, where it is associated with cuprite, limonite, malachite, and azurite. In this association it has been found in Arizona, New Mexico, and northern Mexico.

## PLATINUM, Pt

Platinum is considered with gold a noble metal and, like gold, is not attacked by any of the single acids. Unlike gold, however, it is a relatively new metal, for it was not discovered until 1735. In that year it was found in placer workings in Colombia, South America, and was given the name *platina* from the word *plata*, Spanish for silver, since it was regarded as an impure ore of that metal. One of the first uses for platinum was in counterfeit gold coins. A center of platinum was surrounded with gold, and, because of its high specific gravity, the fraud was difficult to detect. From such a lowly beginning, platinum has taken its place among the most valued metals; it is more valuable in fact than gold itself.

Between the years 1828 and 1845 platinum coins were in circulation in Russia, but they were recalled and the experiment has not been repeated. Today platinum has many uses based mostly on the fact that it is fused with great difficulty and is not attacked by ordinary chemical reagents. It is used in the chemical laboratory for crucibles, dishes, spoons, and other types of laboratory equipment. A few of its many other uses are in electrical apparatus, jewelry, dentistry, and in making photographic prints. For many of its applications it is alloyed with other metals of the platinum group. Thus platinum jewelry commonly contains about 10 per cent of iridium.

**Habit.** Native platinum is isometric but it is rarely found in crystals; its usual occurrence is in small grains and scales. Some nuggets weighing as much as 20 pounds have been found in placer workings in the Ural Mountains.

**Physical Properties.** The hardness is 4–4½, which is high for a metal. Pure platinum, such as is used in the laboratory, has a specific gravity of 21.5, but for native platinum, alloyed with other metals, it is lower, 14–19. It is malleable and ductile but less so than gold, silver, and copper. The color is steel gray. Some platinum rich in iron is magnetic.

**Composition.** Native platinum is found by the chemist to contain considerable amounts of iron and also a number of rare metals of the platinum group, such as iridium, osmium, rhodium, and palladium.

Platinum, like gold, does not readily combine with other elements, the only compound known in nature being an arsenide, $PtAs_2$, called *sperrylite*.

**Occurrence.** Until recently most of the world's platinum has come from placer deposits in the Ural Mountains. Platinum was discovered there in 1822 and, since that time, has been mined continually. The deposits are now becoming exhausted, and an effort is being made to mine the platinum in places where it is

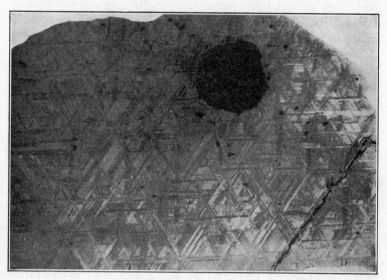

Fig. 194. Iron Meteorite Polished and Etched, Altonah, Utah. The dark patch is troilite, FeS.

found associated with chromite in an olivine rock, peridotite. At present the chief source of platinum is the great ore body at Sudbury, Ontario, where it is recovered from nickel-copper ores. South Africa is also an important producer with large reserves of low-grade platinum ore.

## Iron, Fe

Native iron, or iron occurring in nature in the metallic state, is known only as a great rarity and hence is of no practical importance. The meteorites which occasionally fall to earth often consist entirely of metallic iron; others that have a stony appearance may contain particles of metallic iron distributed through them. The iron of meteorites always contains some nickel. This is important to remember, for many a piece of old iron is found each year that the finder suspects of being a meteorite. A test proving the absence of nickel is proof that the specimen is not a meteorite. If an iron meteorite is polished and etched, a triangular pattern such as is shown in Fig. 194, gives proof of crystallization.

Native iron has been noted also in a few places in terrestrial rocks, but only one occurrence is especially noteworthy—that at Disko, Greenland. Here iron fragments ranging in size from small grains to masses weighing tons have been found in basalts.

The reason that native iron is not found more abundantly is that it oxidizes so easily. Just as an iron implement left outdoors will, after a short while, rust and fall to pieces, so will native iron, for the "rusting" is merely the combining of the iron with oxygen and water.

Iron is our most important industrial metal, and the ores from which it comes, chiefly the oxides or iron, are discussed later. (See hematite, p. 165.)

### Semi-Metals

Of the elements that occur in the native state some, like gold, silver, and copper, are true metals; others, like diamond and graphite, the two forms of carbon, and sulfur are nonmetals. Intermediate between these two groups are tellurium, arsenic, antimony, and bismuth, known as the semi-metals. As minerals they are relatively unimportant, and the sources of these elements are for the most part other minerals.

### Arsenic, As

Native arsenic is usually found in reniform and stalactitic masses. When crystals are found, they are rhombohedral with a perfect basal cleavage. Arsenic has a metallic luster and tin-white color, but it soon tarnishes on the surface to a dull gray. The hardness is 3½; the specific gravity, 5.7. Most of the arsenic used is obtained as a by-product in the smelting of ores for other metals.

### Antimony, Sb

Native antimony usually occurs in massive aggregates which show basal cleavage. It is rhombohedral, tin-white, and brittle.

The hardness is 3–3½, and the specific gravity is 6.7. Native antimony occurs in veins with silver, arsenic, and other antimony minerals. The sulfide, stibnite, is the chief ore of antimony (p. 151).

## BISMUTH, Bi

Of the native semi-metals, bismuth is the most important since it is a major source of that element. The sulfide *bismuthinite* is a rare mineral although some bismuth is obtained from it. Bismuth fuses at a low temperature, and alloys of it with tin, lead, and cadmium fuse at lower temperatures, even below the boiling point of water. For this reason they are used for electric fuses and safety plugs in water sprinkling systems. Much of the bismuth produced is used in medicine and cosmetics.

Native bismuth is rarely in distinct crystals and is usually granular or in small veinlets. Artificial crystals are rhombohedral, pseudocubic. Good basal cleavage is seen even in granular aggregates. The hardness is 2–2½; the specific gravity, 9.8. Bismuth is silver-white with a reddish tinge and has a bright metallic luster. It is sectile but at the same time brittle. Bismuth is a relatively rare mineral and the metal is obtained mostly as a by-product in the smelting of gold and silver ores.

### Nonmetals

## SULFUR, S

Native sulfur is the chief source of that element, and large deposits of it are found in various parts of the world. Not only does sulfur occur in the native state, but also it is present in nature in the sulfides and the sulfates. Although these two groups of minerals contain vast quantities of sulfur, only pyrite, iron sulfide, has been utilized as a source; and the world's supply comes mostly form native sulfur.

Sulfur has many uses, but most of it is consumed in the chemical industry in the manufacture of sulfuric acid. It is used also in fertilizers, insecticides, explosives, paper, and rubber, matches, and gunpowder.

**Habit.** Sulfur has three polymorphic forms. Two are monoclinic and are very rare in nature; the third type, most common as a mineral, is orthorhombic. It is found in crystals usually showing rhombic dipyramids (Figs. 195–197). It occurs also in masses and as fine granular aggregates.

**Physical Properties.** Sulfur is soft with a hardness of $1\frac{1}{2}$–$2\frac{1}{2}$ and, though brittle when struck by a hammer, is easily cut by a knife. The specific gravity is 2.05. It has a resinous luster and a bright yellow color and streak. The color is so characteristic that

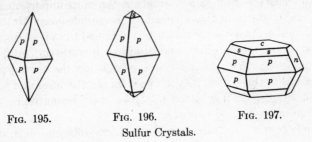

FIG. 195.  FIG. 196.  FIG. 197.

Sulfur Crystals.

it is known as sulfur-yellow. Crystals are often clear and transparent, and some may have a greenish cast. Sulfur is such a poor conductor of heat, that, if a crystal is heated gently by being held in the hand, the surface will expand and spall off. One should therefore not handle good crystals.

**Composition.** Sulfur is usually pure, but it may contain impurities of clay or asphalt. It is remarkable among minerals because when heated it takes fire and burns with a pale blue flame giving off sulfur dioxide which has a very characteristic suffocating odor. This is the best test for sulfur.

**Occurrence.** Sulfur is found on the rims of the craters of some volcanoes as a direct result of volcanic activity and has thus received the old name of *brimstone*. In this form it has been found in Japan and in Chile in large amounts. It is found also in beds associated with gypsum as at the famous locality near Girgenti, Sicily. Most of the world's sulfur today comes from Louisiana and Texas, where it is associated with salt domes. It is not mined but recovered by pumping superheated water down to melt the sulfur in place and then bringing it to the surface by compressed air.

## DIAMOND, C

Of all the minerals described in this book, diamond is probably the most familiar to the average person. Everyone has seen cut diamonds in jewelry, but few have seen diamonds in the rough. It is in the uncut state that they are of the greatest interest to the mineralogist. One of the rarest mineral specimens is a diamond embedded in the rock in which it grew.

**Habit.** Diamonds are usually found in distinct, isolated crystals, most of them very small, but some are as large as English walnuts or even larger. The crystals are usually octahedrons; less commonly other isometric forms, Figs. 198 to 200, are observed.

  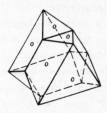

<table>
<tr><td>Fig. 198.</td><td>Fig. 199.</td><td>Fig. 200.</td></tr>
</table>

Diamond Crystals.

The natural crystals frequently have rounded edges and curved faces, or the faces may show little triangular pits like the etchings spoken of on p. 17. There are also crystals with irregular habit, occasionally as round as peas. Twins, such as shown in Fig. 200, also are found. In Brazil a black massive variety known as *carbonado* or *carbon* has been found. It is tougher and less brittle than the crystals.

**Physical Properties.** The hardness of a diamond is 10, greater than that of any other known substance either natural or artificial. Although diamond stands next to corundum in the scale of hardness, it is actually many times harder; the hardness difference between corundum and diamond is greater than that between talc (1 in the scale) and corundum, 9.

The specific gravity is 3.5, which is high for a nonmetallic mineral. The crystals have perfect octahedral cleavage, and, in spite of its high hardness, diamond is brittle because it breaks easily along the cleavage planes. Diamond has an extremely high luster known as *adamantine*. The term is derived from this species and means the brilliancy of the diamond. The brilliancy is increased when a stone is cut with many facets. Many natural crystals have a peculiar greasy appearance. The most highly prized gemstones are colorless and water-clear, but they are not so common as those that are pale yellow in color.  Pale shades of red, orange, green, blue, and brown also are observed.

**Composition.**  Diamond is pure carbon and thus has the same composition as a piece of charcoal.  It is infusible, as is charcoal, and is not acted upon by acids; but, unlike charcoal, it does not burn readily.  When heated very hot in the presence of oxygen, it is slowly consumed, forming, like burning charcoal, carbon dioxide.

**Occurrence.**  All the diamonds that were known from early times until the beginning of the eighteenth century came from India where they were found in stream gravels.  Diamonds were discovered in Brazil in 1729, and since that time India has been of little importance as a producer.  As in India, the Brazilian diamonds are in stream gravels, and the original source from which they came has never been discovered.  Although Brazil is still producing diamonds, its importance declined rapidly after the deposits of South Africa were discovered.

Diamonds were found in gravels of the Vaal River in South Africa in 1868, and since that time the bulk of the world's supply has come from the African continent.  Shortly after their discovery in stream gravels, diamonds were found in what later were called "diamond pipes."  Here the diamonds were embedded in a rock, called *kimberlite*, that was yellow at the surface and blue in depth.  The kimberlite was thus given the names "yellow gound" and "blue ground."  This rock, unlike the stream gravels, was the place that the diamond actually crystallized.  These great "pipes" extend thousands of feet in depth and contain diamonds uniformly but sparsely distributed through them.  It is estimated that of the material that makes up the "pipes" 1 part in 14 million is diamond!  Several diamond pipes were discovered in South Africa during the latter part of the nineteenth century, and in 1902 the largest of all, the Premier mine, was found in the Transvaal.  More recently other "pipes" have been found in Rhodesia.  Diamonds have been found in stream gravels in French Angola, the Gold Coast, French Equatorial Africa, Tanganyika, and, most important of all, the Belgian Congo.  During the war years of 1940–1945 the Belgian Congo was the world's chief producer, particularly of industrial diamonds.

Diamonds have been reported from many parts of the United States, but only as isolated stones here and there.  The only place in the United States that can be called a diamond deposit is in Arkansas, where in 1906 a diamond "pipe" similar to those of Africa was discovered. Mining there has produced some 40,000 stones, but for many years, because of the high cost of operation, it has been unproductive.

**Uses.** Of all the diamonds that are mined, only about 20 per cent are suitable for cutting into gemstones; the others are used for industrial purposes. The industrial uses for the off-colored or flawed stones, known as *bort*, are many. Drills set with diamonds are used by the miner to obtain samples of rock in advance of his workings to tell him the extent and grade of the ore. Larger diamond drills are used for drilling thousands of feet through solid rock. Metal discs impregnated with diamonds are used for sawing rocks and other hard material. Diamond dies are used for drawing most fine wire. For this purpose a hole is drilled through the diamond and the wire is drawn through it thus reducing the wire to the diameter of the hole. Diamonds are also used for cutting glass and, in the form of powder, in grinding diamonds and other hard gems.

Everyone knows the use of the diamond for jewelry, for which its brilliancy, hardness, and comparative rarity make it the most important of all gemstones. The clear, colorless or "blue-white" stones are in general the most valuable; a faint yellow color, often present, detracts from their value. However, deep shades of yellow, red, green, or blue greatly increase the value.

## GRAPHITE, C

The name graphite comes from the Greek meaning *to write* because of its use in pencils and crayons. It has also been called *plumbago* and *black lead*. Both these names were given because it was confused with galena, the common black lead sulfide. Only in its black color does it resemble galena, however; the other properties of both minerals are quite distinct.

It is interesting to compare the two natural forms of carbon— diamond and graphite—for it is difficult to picture two substances of like composition with such vastly different properties. Diamond is the hardest of minerals, light in color, and high in specific gravity; graphite is one of the softest of minerals, black and opaque, with a low specific gravity. It is the difference in the arrangement of the atoms in the two minerals that gives rise to the different properties.

**Habit.**  Graphite is hexagonal, and some crystals have a hexagonal outline with a prominent basal plane.  Usually it appears massive but may be separated easily into thin leaves or plates; hence it is said to be foliated.  It may also be finely granular and compact.

**Physical Properties.**  Graphite is sectile and so soft (its hardness is 1–2) that it makes a mark on paper and feels greasy to the hand.  Its specific gravity is 2.2, very low for a mineral with a metallic luster.  The color is iron-black to steel-gray.  A good cleavage parallel to the base permits the easy separation of the plates.

**Composition.**  Graphite is pure carbon, but some deposits may be impure, mixed with clay, iron oxide, or other minerals.

**Occurrence.**  Graphite is commonly found in small scales scattered through rocks such as marbles, crystalline limestones, schists, and gneisses.  Occasionally it is in large beds or veins that can be mined.  At Sonora, Mexico, an igneous rock has intruded a coal bed and converted adjacent coal to graphite.  It is mined most extensively on the Island of Ceylon, but some mining is done in Austria, Italy, India, and Mexico.  In the United States the eastern part of the Adirondack region, particularly at Ticonderoga, has been most productive.

To the beginning of the twentieth century all the graphite used in industry was the natural mineral and was mined.  Since that time graphite has been manufactured by heating coke to an intense heat in an electric furnace.  At present more synthetic than natural graphite is used.

**Uses.**  Graphite mixed with clay is the so-called *lead* of our "lead pencils."  It is also mixed with clay for making crucibles for handling molten metal, because it is infusible, is not affected by the heat of an ordinary furnace, and will not react with the molten metal.  It is an excellent lubricant and as such is frequently mixed with oil.  It is also used in electroplating, foundry facings, stove polish, and in many other ways.

## SULFIDES

The sulfides form an important group of minerals, for among them are many of the ores of the common metals.  In them sulfur is combined with one or more metals, and by heating the mineral before the blowpipe the sulfur unites with oxygen and is driven off as sulfur dioxide gas. It can be detected by its pungent and irritating odor. Included with the sulfides are several tellurides, that is, those compounds in which tellurium instead of sulfur has combined with a metal.

In a more extensive treatment of mineralogy than is given in this book, the sulfides would be followed by a group of minerals called the *sulfosalts*. This group is fairly large, but only a few of the minerals belonging to it are common enough to warrant description here. Consequently, after the tellurides but without separate grouping, the four most important sulfosalts, pyrargyrite, proustite, tetrahedrite, and enargite, are described.

## ARGENTITE, $Ag_2S$

Argentite, sometimes called *silver glance*, is named from the Latin word for silver, *argentum*. It is not very common, but it is an extremely valuable ore, since when pure it contains 87 per cent of metallic silver.

**Habit.** Argentite is isometric, and crystals are usually cubic or octahedral. Most commonly it is massive or in coatings on other minerals.

**Physical Properties.** The hardness is about 2, and the specific gravity, 7.3. It is eminently sectile and can be cut with a knife almost as easily as lead. It can be flattened to some extent under the hammer, whereas almost all other sulfides being brittle break into many fragments. The luster is metallic, and the color and streak grayish black.

**Composition.** The formula is $Ag_2S$. Heated by the blowpipe flame on charcoal, the sulfur is easily roasted off and a little silver ball is left behind. The ball can be tested chemically by dissolving it in nitric acid and adding a drop of hydrochloric acid. A white, curdy precipitate of silver chloride will be formed.

**Occurrence.** Argentite is found in veins associated with other silver minerals, and it may be found also near the surface as the alteration product of other minerals. It is an important ore of silver at various localities in Mexico, Peru, Chile, and Bolivia. In the United States it has been mined as a valuable ore at Virginia City and Tonopah, Nevada.

## CHALCOCITE, $Cu_2S$

Chalcocite, or *copper glance*, is one of the most valuable ores of copper, for when pure it contains about 80 per cent of the metal. If one reads the history of copper mining, he finds that first one mineral and then another was the chief copper ore. For example, during the middle part of the nineteenth century, native copper from the Lake Superior district was the most important in the United States. Today chalcocite is most important, and much of it is found in tiny grains scattered through large bodies of rock called porphyry copper ores. The mining of these ores is carried out on a gigantic scale, for a great deal of rock must be mined to obtain a pound of copper.

**Habit.** Chalcocite is orthorhombic, but crystals are very rare and usually small. It is most commonly fine-grained and massive.

**Physical Properties.** The hardness is $2\frac{1}{2}$–3, and the specific gravity 5.6. It is brittle when struck with a hammer but can be cut to some degree with a knife. It is this property of being imperfectly sectile by which miners sometimes identify it. It is black or bluish black and when fresh has a brilliant metallic luster. The streak is grayish black. When exposed to the air it becomes dull and tarnished on the surface.

**Composition.** The formula is $Cu_2S$. When heated on charcoal with the blowpipe flame the sulfur is easily roasted off, leaving behind a little ball of metallic copper.

Another, less common copper sulfide contains only 66.4 per cent of copper with the formula CuS. This is *covellite*. Like chalcocite it forms as a secondary mineral, but it is easily distinguished from chalcocite because of its indigo-blue color.

**Occurrence.** Most chalcocite is secondary, that is, it was formed by the alteration of some earlier copper mineral by solutions working downward from the surface. In such a way large masses of solid massive chalcocite have formed, such as those at Clifton and Morence, Arizona. Elsewhere, as at Bingham, Utah, this process has enriched other sulfides disseminated through large masses of rock by replacing them with chalcocite to form workable deposits called "porphyry copper" ore. Such deposits rarely contain more than 2 or 3 per cent of copper, but because they can be worked on such a large scale they furnish most of the copper produced in the United States today. Chalcocite is an important ore mineral at the great copper deposit at Butte, Montana.

Crystals of chalcocite are rare, but fine specimens have in the past come from Cornwall, England, and Bristol, Connecticut.

## BORNITE, $Cu_5FeS_4$

Bornite was named after the Austrian mineralogist von Born but it has a variety of other names, such as *purple copper ore*, *variegated copper ore, peacock ore*. All these suggest a property by which it is easily recognized: the bright iridescent tarnish of the surface.

**Habit.** Bornite is isometric and at a few places has been found in rough cubic crystals, but usually it is massive as embedded particles or irregular grains.

**Physical Properties.** The variegated purple to blue color on the exposed surface of bornite is its most characteristic feature. However, exposed surfaces of other copper minerals may be somewhat similar, so that, in identifying a specimen, one should always break off a small fragment and observe the fresh fracture, which is brownish bronze in color. The hardness is 3; the specific gravity 5.07. The luster is metallic, and the streak grayish black.

**Composition.** Bornite contains both copper and iron and has the formula $Cu_5FeS_4$. When heated with the blowpipe flame on charcoal, the sulfur is driven off and the fragment becomes magnetic, showing the presence of iron. If the fragment is touched with a drop of hydrochloric acid and is again heated, the azure-blue flame of copper chloride shows the presence of copper.

**Occurrence.** Bornite is a widespread and important ore of copper, usually associated with other copper sulfides. At most of the copper mines of western United States bornite is found to some extent, but it is a particularly important mineral at Superior, Arizona; Butte, Montana; Magma Mine, Arizona; and Engles Mine, California.

## GALENA, PbS

Galena is the most important ore of lead and one of the commonest minerals. Other ores of lead, such as the sulfate, anglesite, and the carbonate, cerussite, will be considered later. *Native lead* is a very rare mineral, though occasionally found in small amounts.

Lead is one of the most important metals, used for many purposes familiar to all, such as drain pipes, storage batteries, cable coverings, foil, bullets, and shot. It is used also in solder alloyed with tin and in type metal alloyed with antimony. Much lead is made into the basic carbonate, *white lead*, and is used as a paint pigment.

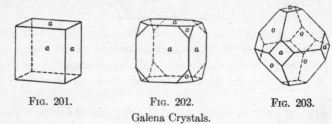

Fig. 201.       Fig. 202.       Fig. 203.

Galena Crystals.

**Habit.** Galena crystallizes in the isometric system, and well-formed cubic crystals are common. It is found also in octahedrons

Fig. 204.   Galena Crystals, Breckenridge, Colorado.

or in crystals showing a combination of cube and octahedron (Fig. 203).

**Physical Properties.** A perfect cubic cleavage is the most outstanding feature by which galena can be identified. A fragment struck with a hammer will be seen to break up into a multitude of rectangular blocks. Even when a specimen of galena is fine-grained and appears almost massive, the eye can catch the light reflected from the myriad tiny cleavage faces. The hardness is 2½, and the specific gravity is 7.5, nearly as high as metallic iron. Lead is a metal of such high density (specific gravity 11.4) that all its compounds are heavy. The luster of galena is metallic and usually very brilliant; the color and streak are bluish lead-gray, but the exposed surface may be somewhat dull from tarnish.

Fig. 205. Galena and Sphalerite on Dolomite, Joplin, Missouri.

**Composition.** Galena is lead sulfide, PbS, but it may contain small amounts of impurities. One of the most interesting of these is silver probably in the form of the sulfide. When it is present in quantity sufficient to justify its being worked for the precious metal, galena is regarded as a silver ore and is called *argentiferous galena*.

When galena is heated on charcoal before the blowpipe, it fuses easily and on continued heating will yield a globule of metallic lead.

**Occurrence.** Wherever lead is mined extensively anywhere in the world, galena is the chief ore mineral. In the United States some of the most important regions are: southeastern Missouri; the tristate district of Missouri, Kansas, and Oklahoma; the Coeur d'Alene district, Idaho; and the Leadville district, Colorado.

Sphalerite, pyrite, hemimorphite, smithsonite, and chalcopyrite are common accompanying metallic minerals. Quartz, calcite, barite, and fluorite are common nonmetallic minerals associated with it. As the result of its own decomposition, cerussite (lead carbonate) and anglesite (lead sulfate) are often found with galena.

### SPHALERITE, ZnS

Sphalerite is named from a Greek work meaning *treacherous*. Certain varieties of it are hard to identify, and the young mineralogist, after misidentifying it several times, will think it well-named, because it often occurs with and is mistaken for the more easily recognized lead ore, galena. The miner's names *black jack* and *false galena* refer to the same fact. Another name is *zinc blende*, chosen, for the same reason, from the German word meaning *blind* or *deceiving*.

FIG. 206.                                FIG. 207.

Sphalerite Crystals.

Sphalerite is the most important ore of zinc, a useful metal in industry and the arts. Iron in sheets and in wire is frequently coated with zinc to protect it from rusting and is said to be *galvanized*. Zinc is used in storage and dry-cell batteries, and alloyed with copper it forms brass. Large amounts are used in the pigments zinc oxide and lithopone (a mixture of zinc sulfate and barium sulfide).

**Habit.** Sphalerite is isometric and when it is well-crystallized is found in tetrahedrons. The crystals, however, are usually twinned and in distorted aggregates, and it requires a trained and skillful eye to understand them. It is usually found in coarse- to fine-granular, cleavable masses.

**Physical Properties.** Sphalerite has perfect dodecahedral cleavage; from coarse masses it is possible to cleave an almost perfect dodecahedron. Even if the sphalerite is fine-granular, the cleavage surfaces are usually prominent, though some types are so closely compacted that they show no cleavage.

The hardness is 3½–4; the specific gravity, about 4. In the rare perfectly pure specimen sphalerite is clear and nearly colorless and has an adamantine luster. Commonly it contains some iron, and the color deepens with increasing amounts of iron from yellow to yellowish brown (the most common) to black. Some crystals known as *ruby zinc* are red.

The luster in most specimens of sphalerite is resinous and is so distinct that the mineralogist comes to depend upon it to enable him to identify the mineral. The streak is white, pale yellow, or brownish, becoming deeper the darker the color of the mass, but always lighter in color than the massive mineral.

**Composition.** Sphalerite is zinc sulfide, ZnS, but, as stated above, iron is usually present, and black sphalerite may contain as much as 18 per cent. Manganese and the rare element cadmium are usually present in small amounts. In some places the rare mineral *greenockite*, cadmium sulfide, is found in earthy crusts on sphalerite.

The tests for sphalerite are poor. Before the blowpipe it does not fuse, but, if powdered, mixed with sodium carbonate, and heated on charcoal, it gives a zinc coating that is yellow when hot and white when cold. When warmed in a test tube with hydrochloric acid, it effervesces, giving off bubbles of gas that one might mistake for carbon dioxide, except that the disagreeable odor shows it to be hydrogen sulfide, $H_2S$.

**Occurrence.** Sphalerite is one of the commonest metallic minerals and is frequently associated with galena, chalcopyrite, pyrite, and other sulfides. It occurs both in veins and in irregular deposits in limestone. In limestone deposits it is found in the Joplin district in southwestern Missouri and in adjacent portions of Kansas and Oklahoma. Other large zinc deposits are in Colorado, Montana, Wisconsin, and Idaho.

## CHALCOPYRITE, $CuFeS_2$

Chalcopyrite, or *copper pyrites*, as it is sometimes called, is an important ore of copper. The same elements are present as in bornite but in different proportions. The color of chalcopyrite is a beautiful deep brass-yellow, so golden that chalcopyrite is frequently mistaken for gold, especially when it is scattered in small particles through a mass of quartz. Although it can be easily distinguished from gold as we shall see, the name "fool's gold," which it shares with pyrite, is not inappropriate.

It is interesting to compare the percentages of copper in the four copper minerals considered thus far; native copper, 100%; chalcocite, 79.8%; bornite, 63.3%; chalcopyrite, 34.5%.

Fig. 208.                     Fig. 209.

Chalcopyrite Crystals.

**Habit.** Chalcopyrite belongs to a low symmetry class of the tetragonal system, and crystals are usually sphenoids. The crystals appear isometric, and it is difficult with the unaided eye to distinguish them from tetrahedrons. Usually chalcopyrite is massive and as such may be found in large specimens or in tiny specks in the enclosing rock.

**Physical Properties.** The hardness of chalcopyrite is $3\frac{1}{2}$–4, and the specific gravity is a little over 4. The luster is brilliant metallic, and the color, as we have seen, deep brass-yellow; the streak is greenish black. A tarnish often develops on the surface deepening the color or giving it a variegated appearance similar to bornite. To determine the color one should always examine a fresh fracture.

Chalcopyrite can be readily distinguished from pyrite, for, unlike pyrite, it can be easily scratched with a knife and its color is deeper. It is distinguished from gold by being brittle. When scratched with a knife, it breaks into fragments, whereas the scratch in gold is a shiny groove.

**Composition.** The formula is $CuFeS_2$. Heated on charcoal, a fragment fuses to a black mass that is strongly magnetic. Dissolved in nitric acid, it gives a blue solution that turns azure-blue when an excess of ammonia is added.

**Occurrence.** Chalcopyrite is a very common mineral, the most widespread of the copper sulfides, and one can expect to find small amounts of it almost anywhere. It frequently occurs in veins associated with galena, sphalerite, and pyrite, as well as other copper minerals. A few of the localities at which it is an important ore of copper are Butte, Montana; Bingham, Utah; and Jerome, Arizona.

## PYRRHOTITE, $Fe_{1-x}S$

Pyrrhotite takes its name from a Greek word meaning *reddish* because of its peculiar reddish bronze color; its color is a very important property to remember as an aid in its identification. It is called also *magnetic pyrites*, which refers to its being magnetic, a still more striking property. Other sulfides containing iron are magnetic after heating, but pyrrhotite alone is attracted by the magnet without heating.

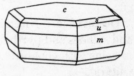

FIG. 210. Pyrrhotite.

**Habit.** Pyrrhotite crystallizes in the hexagonal system, but good crystals are rare. When found, they are tabular showing a hexagonal outline. It is usually found in irregular masses.

**Physical Properties.** The hardness is 4, and the specific gravity 4.6. The luster is metallic, and the color, as noted before, a peculiar reddish bronze quite different from pyrite and marcasite. The streak is black.

**Composition.** Pyrrhotite is a sulfide of iron expressed by the unusual formula $Fe_{1-x}S$, where $x$ lies between 0 and 0.2. This means that it is not quite equivalent to the simple sulfide FeS, where Fe and S are in the ratio of 1 : 1; but there is a deficiency in iron that differs in different specimens.

**Occurrence.** Pyrrhotite is a common minor constituent of igneous rocks, and in certain basic igneous rocks it occurs in large masses associated with chalcopyrite and pentlandite, (Fe,Ni)S. Because of the associated nickel-bearing pentlandite, pyrrhotite is mined on a large scale and can be considered the world's chief nickel ore. As such it is mined at the great nickel mines at Sudbury, Ontario. It is of no value as an iron ore.

## NICCOLITE, NiAs

Niccolite is often called *copper nickel* which comes from the German *kupfernickel*, the first name given to the mineral. It is called copper nickel only because of its conspicuous copper color, for it contains no copper but is a minor ore of nickel.

**Habit.** It is rarely in hexagonal crystals and is usually massive or reniform with columnar habit.

**Physical Properties.** The hardness is 5–5½, and the specific gravity is 7.8. The color, as already noted, is pale copper-red; the streak is brownish black.

**Composition.** Niccolite is nickel arsenide, NiAs. It is rarely pure and usually contains a little iron, cobalt, sulfur, and antimony. When it is heated on charcoal, dense white fumes of arsenious oxide form and the characteristic garliclike odor of arsenic is given off.

**Occurrence.** Niccolite is associated with other nickel arsenides and sulfides and pyrrhotite in basic igneous rocks. It is found also in veins with cobaltite, smaltite, and silver minerals, as at Cobalt, Ontario.

## MILLERITE, NiS

Millerite is remarkable among minerals because of its occurrence in very fine hairlike or capillary crystals, and for this reason it is called *capillary pyrites*.

**Habit.** Millerite is rhombohedral, and the crystals are usually greatly elongated parallel to the *c* crystal axis forming, in places, masses that resemble a wad of hair. In other localities delicate, radiating crystals or crusts with a fibrous habit are found in cavities.

**Physical Properties.** The hardness is 3–3½; the specific gravity, 5.5. The luster is metallic, and the color, pale brass-yellow with a greenish tinge when the crystals are in fine hairlike masses. The streak is greenish black.

**Composition.** Millerite is nickel sulfide, NiS. When it is heated before the blowpipe, the sulfur is driven off and the remaining fragment, after heating in the reducing flame, is attracted by a magnet. It should be remembered that nickel, like iron, is magnetic, though to a lesser intensity.

**Occurrence.** As stated before, millerite occurs in hairlike crystals, but in places, as in the geodes in the St. Louis limestone, they are matted together like a wad of hair. At Antwerp, New York, they are found as tufts of extremely delicate radiating crystals in cavities of hematite. At the Gap Mine, Lancaster County, Pennsylvania, they have a compact fibrous habit. In some places millerite fibers are found as inclusions in other minerals.

Only rarely is it found abundantly enough to be an ore of nickel. The iron-nickel sulfide *pentlandite* is more important.

## CINNABAR, HgS

Cinnabar is the source of the world's supply of mercury, or *quicksilver*, and is thus a very important mineral. If pure, it is easily recognized by its extremely high specific gravity and its red color.

Native mercury has been found in small amounts associated with cinnabar, but such occurrences are rare. Mercury is the only metal that is liquid at ordinary temperatures, for it becomes a solid only when cooled to $-39°$ F. This property gives it many uses. We are familiar with it in clinical and household thermometers. It is used also in barometers, pressure gauges, electrical switches, and various other scientific apparatus. When the dentist puts a "silver" filling in a tooth, he mixes mercury with silver to form an amalgam. One of the methods of recovering gold and silver is to use mercury, which forms an amalgam with these metals. Other important uses are in the manufacture of fulminate of mercury for detonating high explosives and in the preparation of drugs.

**Habit.** Cinnabar is hexagonal but is usually found in fine granular masses or disseminated through the rock in which it occurs. Crystals that are occasionally found are usually rhombohedral or prismatic.

**Physical Properties.** Cinnabar has perfect prismatic cleavage, which gives granular aggregates a sparkle as the light is reflected from the many brilliant faces. The hardness is $2\frac{1}{2}$; the specific gravity, 8.1, higher than that of metallic iron (7.8). The great weight cannot escape the observer and is a striking characteristic. In some specimens, however, the cinnabar is not pure but is scattered through a clayey gangue and gives its color to the rock. In such cases the density of the whole is much lower. The color is vermilion-red; the streak, scarlet; the luster, adamantine.

**Composition.** Mercury sulfide, HgS. Because of the high atomic weight of Hg, cinnabar contains 86.2 per cent of mercury. If heated on charcoal, a piece of pure cinnabar is volatilized entirely. In an open tube, if it is heated very slowly, so that the sulfur has time to oxidize, a ring of metallic mercury is formed on the cold part of the tube. If the cinnabar is heated too rapidly, a ring of black mercury sulfide will form that has the same composition as the original mineral. (See p. 117.)

**Occurrence.** Cinnabar is found at many localities, but at only a few of them is it in sufficient quantity to be mined. The most important localities are at Almaden, Spain; and Idria, Italy. It has been mined also at New Almaden and New Idria, California, as well as at various places in Oregon, Arkansas, Texas, and Nevada.

## REALGAR, AsS; ORPIMENT, $As_2S_3$

Realgar and orpiment can be considered together, since both are compounds of arsenic and sulfur and are almost invariably associated with each other.

REALGAR is found in transparent monoclinic crystals and massive aggregates which have a beautiful aurora-red color. It is soft and sectile (hardness is $1\frac{1}{2}$–2) and has a specific gravity of 3.5. The luster is resinous, and the streak red to orange.

Realgar is arsenic monosulfide, AsS. It is easily fusible and, when heated on charcoal, gives a volatile white sublimate of arsenious oxide with the characteristic garlic odor.

ORPIMENT, named from the Latin *auripigmentum*, meaning *golden paint*, is a beautiful golden yellow. It is monoclinic, but distinct crystals are rare, and it is generally found in foliated or columnar masses. It has a perfect side pinacoid cleavage so that it splits easily into thin flexible leaves. The luster is pearly on this cleavage face. It is soft and sectile (the hardness $1\frac{1}{2}$–2). The specific gravity is 3.5.

The composition of orpiment is $As_2S_3$, or arsenic trisulfide. When it is heated on charcoal, its behavior is like that of realgar; it volatilizes completely and gives the characteristic garlic odor of arsenic and white fumes of the oxide, $As_2O_3$.

**Occurrence.** Realgar and orpiment are found together, frequently with other arsenic minerals and stibnite in veins of lead, silver, and gold ores. They are deposited also from geyser water as at Yellowstone National Park. Mercur, Utah, and Manhattan, Nevada, are two of the most important United States localities.

## STIBNITE, Sb$_2$S$_3$

Stibnite, sometimes called *antimony glance*, is the commonest and most important ore of antimony, a metal that is very useful in the arts. One part of antimony is alloyed with 3 parts of lead to form type metal. Antimony has the unusual property of expanding on cooling and causes the alloy to fill out the mould and give sharp clean letters. Babbitt metal, used in bearings, is an alloy of antimony with tin and copper.

**Habit.** Stibnite is orthorhombic and is frequently found in prismatic crystals vertically striated, often with spear-shaped terminations (Fig. 211). Some crystals may be curved or bent. Stibnite is found also in radiating groups as well as granular aggregates.

**Physical Properties.** There is perfect cleavage parallel to the side pinacoid, and the cleavage surfaces are smooth and highly polished, sometimes showing cross striations. Even in the granular aggregates, the cleavage is usually visible. The hardness is 2, so that it is easily scratched and will leave a mark on paper. It should not be confused with graphite, which is much softer and greasy to the touch and marks paper without the slightest tendency to tear it. The specific gravity is 4.6. The luster is metallic and, on a fresh cleavage surface, is very brilliant. The color and streak are lead-gray to black.

FIG. 211. Stibnite.

**Composition.** Stibnite is antimony trisulfide, Sb$_2$S$_3$. It fuses very easily in a match flame (1 in the scale of fusibility). When heated on charcoal, it gives off fumes of the oxide, Sb$_2$O$_3$, which form a thick coating at a little distance; after a few moments, the fragment is entirely volatilized. If the reducing flame is thrown for a moment on the coating, it is burned off with a greenish blue flame. (See p. 116.)

**Occurrence.** Stibnite is associated with other antimony minerals that are the product of its decomposition; it is also frequently found with realgar, orpiment, cinnabar, sphalerite, barite, and galena. Most of the world's supply comes from China, but the

most perfect and beautiful crystals have come from Japan.   In
the United States the chief occurrences are in Nevada, California,
and Idaho.

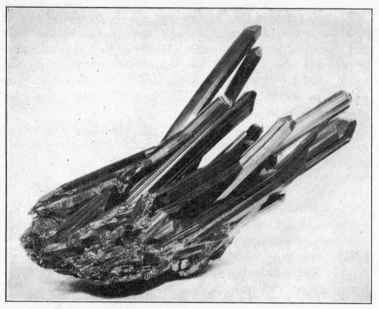

FIG. 212.   Stibnite Crystals, Japan.

## PYRITE,  $FeS_2$

Pyrite, or *iron pyrites*, is one of the minerals known as "fool's
gold."  It is the most common, as well as the most striking,
sulfide mineral.  It forms under an extremely wide range of
conditions and is thus associated with many minerals, most commonly with chalcopyrite, sphalerite, and galena.

Massive pyrite is mined in some places for the gold and copper
associated with it but only under unusual circumstances for its
constituent elements alone.  When thus mined, it is usually for
the sulfur rather than the iron, for sulfur makes up 53.4 per cent
of pyrite.  In a few countries where there are none of the richer
oxide ores, it has been mined on a small scale as an ore of iron.

**Habit.**   Pyrite is often found in cubic crystals the faces of which
usually show fine lines or striations parallel to one pair of edges
only (Fig. 213) and at right angles to those on the adjoining faces.

These striations are the result of an oscillatory combination of the cube faces with those of the pyritohedron. The pyritohedron (Fig. 214), named from this species, is also common; and it, like the cube, may show fine striations. The two forms may appear together more or less equally well-developed (Fig. 215). More rarely one may find octahedrons of pyrite, but this form is more common in combination with the cube or the pyritohedron (Figs. 216, 217).

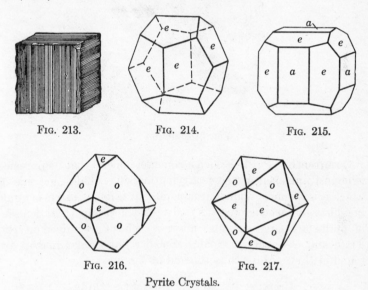

FIG. 213.      FIG. 214.      FIG. 215.

FIG. 216.      FIG. 217.

Pyrite Crystals.

Pyrite is also found in massive or granular aggregates which in some places form huge lenses that may be mined.

**Physical Properties.** The hardness of pyrite is a little over 6, so that it scratches glass and is not scratched by the knife blade. Its hardness, therefore, high for a sulfide, distinguishes it from chalcopyrite (easily scratched by the knife) with which it may be confused. The specific gravity is 5. The luster is brilliant, metallic; and the color, light brass-yellow, growing a little darker when tarnished. The streak is greenish black.

**Composition.** The composition of pyrite is iron disulfide, $FeS_2$. As mentioned previously, small amounts of copper and gold may be present as impurities. When these occur pyrite is mined for the small amount of these metals that can be recovered when it is smelted. Before the blowpipe it fuses on charcoal to a black magnetic bead and gives off abundant sulfur which burns, producing the suffocating fumes of sulfur dioxide.

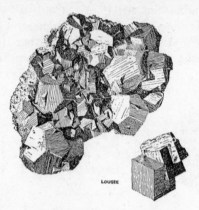

FIG. 218.   Pyrite.

**Occurrence.**   Pyrite is such a common mineral that its presence is almost universal in veins carrying metallic sulfides, in some of which it is the most abundant mineral.   It is found also in crystals in schists and in concretions in coal.   The most important deposits of pyrite are large granular masses, such as are mined at Rio Tinto and elsewhere in Spain.   Smaller but similar masses are found in Rowe, Massachusetts, and in Virginia.

### COBALTITE, CoAsS; SMALTITE, CoAs$_2$

Cobalt is a rare element produced chiefly from cobaltite and smaltite.   The mineral *erythrite*, or cobalt bloom, is a bright rose-red arsenate of cobalt that forms as an oxidation product on cobaltite and smaltite.   Until relatively recent years cobalt had little use in the arts but was chiefly used as *smalt*, a cobalt glass, which has a beautiful blue ultramarine color, and ground up was used as a pigment in coloring glass and pottery.   Today, however, metallic cobalt is alloyed with chromium and molybdenum or tungsten for the manufacture of instruments which must be noncorrosive and yet maintain a keen cutting edge.

COBALTITE is isometric and commonly is found in cubes or pyritohedrons, thus resembling pyrite; but it is distinguished from that mineral by its silver-white color and a perfect cubic cleavage.

The hardness is 5½, and the specific gravity is 6.33. The composition is CoAsS, but it usually contains some iron and nickel. When heated on charcoal it gives white arsenious oxide with the characteristic garlic odor.

SMALTITE, like cobaltite, is isometric and silver-white in color. However, unlike cobaltite, it lacks a cleavage and is rarely in crystals, but it is usually massive. The hardness is 5½–6, and the specific gravity is 6.5. Smaltite is cobalt diarsenide, $CoAs_2$, but it frequently contains some nickel and sulfur.

**Occurrence.** Cobaltite and smaltite occur together usually associated with niccolite, native silver, bismuth, arsenopyrite, and calcite.

Such an association was found both in Saxony and at Cobalt, Ontario. Today much of the world's cobalt comes from the Belgian Congo.

## MARCASITE, $FeS_2$

Marcasite, or *white iron pyrites*, as it is sometimes called, has the same chemical composition as the more common mineral pyrite. It differs from it, however, in its crystal form and physical properties. The two are consequently distinct minerals, and the compound $FeS_2$ is said to be *dimorphous* (see p. 99).

**Habit.** Marcasite is orthorhombic and is commonly in tabular crystals flattened parallel to the basal plane. It is often twinned, and to the grouping that results the fanciful names of *spear pyrites* (Fig. 219) and *cockscomb pyrites* have been given. The crystal habit of marcasite is so characteristic that it is usually easy to distinguish the crystals from the cubes and pyritohedrons of pyrite. It is also found in nodules, in spherical masses, and as stalactites.

**Physical Properties.** The hardness is 6–6½, or about the same as that of pyrite, but the specific gravity is lower, 4.9 instead of 5.0. Unless the specimen is very pure this difference may not be easily detected, but a distinction can usually be made on the basis of color. The name *white iron pyrites* has been given to marcasite because on fresh surfaces it is a paler yellow than pyrite, a difference that becomes quite apparent after comparing a few specimens. It tarnishes easily, however, and one should be certain that a fresh surface is being examined.

**Composition.**   The composition is iron disulfide, $FeS_2$, the same as pyrite, but it tends to disintegrate more easily and some specimens fall to pieces after a short exposure to the air.

**Occurrence.**   Marcasite is much less common than pyrite and forms under a much narrower range of conditions.   It is believed to be a near-surface, low-temperature mineral and forms in concretions in clays and shales and as replacements in limestone.

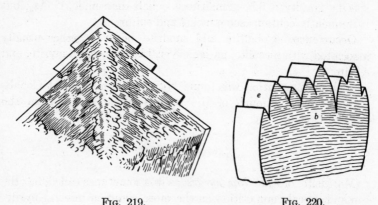

FIG. 219.                    FIG. 220.

Cockscomb Marcasite.

### ARSENOPYRITE,  FeAsS

Arsenopyrite, like pyrite, contains iron and sulfur but, as it contains arsenic as well, it was formerly called *arsenical pyrites*. *Mispickel* is another name that has been applied to it.   It is the most common arsenic-bearing mineral and thus is the chief ore of that element, although much arsenic is obtained as a by-product in the smelting of copper ores.   Arsenic is chiefly used to make the white arsenious oxide which is used as a poison, a preservative, and a pigment.

**Habit.**   Arsenopyrite is monoclinic, but the small crystals in which it is sometimes found have a definite orthorhombic symmetry (Figs. 221 to 223) produced by twinning.   Another type of twinning produces groups which strongly resemble those of marcasite.   It occurs most commonly massive.

**Physical Properties.**   The hardness is 5½–6, and the specific gravity 6.1.   The color is silver-white when fresh, becoming a little dull and tarnished after exposure.   It is much the commonest of the silver-white minerals, and its color readily distinguishes it from the bronze of pyrrhotite or the pale yellow of pyrite and marcasite. The luster is metallic, and the streak grayish black.

**Composition.** Iron arsenide sulfide, FeAsS. When cobalt takes the place of part of the iron, the name *danaite* has been given to the modification. Before the blowpipe it fuses on charcoal to a magnetic globule and gives a coating of white arsenious oxide and the characteristic garlic odor.

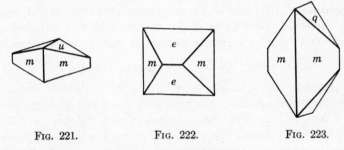

FIG. 221.     FIG. 222.     FIG. 223.

Arsenopyrite Crystals.

**Occurrence.** Arsenopyrite is a widespread mineral formed under varying conditions of deposition. It is frequently associated with ores of tin, tungsten, silver, and copper. Some arsenopyrite is auriferous and is mined for the gold it yields, as at Lead, South Dakota, and Deloro, Ontario.

## MOLYBDENITE, $MoS_2$

Molybdenite is an important mineral for it is the only commercial source of the relatively rare element molybdenum. Today this element has an important place as an alloy in high-speed tool steels. It is used with tungsten to increase the toughness of steel and to permit steel to maintain its cutting edge at a high temperature.

**Habit.** Molybdenite is hexagonal and is rarely found in distinct hexagonal crystals. It is usually in foliated masses or scales.

**Physical Properties.** In most of its physical properties molybdenite resembles graphite, with which it is easily confused. It has a perfect basal cleavage. It is very soft (the hardness is $1-1\frac{1}{2}$) with a soapy feel and leaves a trace on paper. The color is black with a bluish tone, and the specific gravity (4.7) is higher than that of graphite. It is sectile with a metallic luster. The streak is black when observed in the ordinary manner, but on glazed porcelain it is greenish whereas that of graphite is black.

**Composition.** Molybdenum sulfide, $MoS_2$. Heated on charcoal it gives off strong sulfur fumes and yields a deposit of molybdic oxide, which is pale yellow or white. This test further distinguishes it from graphite, for graphite gives no reaction on charcoal.

**Occurrence.** Molybdenite is not a common mineral but is found in many localities as a minor constituent in granites and pegmatites. It is found also in veins associated with tin and tungsten minerals. At Climax, Colorado, molybdenite is found in quartz veins cutting a granite and is here mined on a large scale.

## CALAVARITE, $AuTe_2$; SYLVANITE, $(Au,Ag)Te_2$

Calavarite and sylvanite until late in the nineteenth century were very rare minerals, but with their discovery at Kalgoorlie, West Australia, in 1886, and at Cripple Creek, Colorado, in 1891, they became extremely important. They have been the ore minerals at these two localities that have yielded well over $1,000,000,000 in gold. Both places, however, have been largely worked out, and their present yield is small.

CALAVARITE is monoclinic but is rarely in good crystals and is usually massive. Its hardness is $2\frac{1}{2}$, and its specific gravity is 9.35. The luster is metallic, and the color brass-yellow to silver-white. Its composition is $AuTe_2$, which yields 44 per cent gold. It is easily fusible on charcoal with a bluish green flame, and, after the tellurium is thus driven off, a metallic globule of gold remains.

SYLVANITE is monoclinic, but distinct crystals are rare. It has been called *graphic tellurium* because of the skeleton forms, resembling written characters, that the crystals sometime take on a rock surface. Unlike calavarite, it has one good direction of cleavage. The hardness is $1\frac{1}{2}$–2, and the specific gravity is 8.0–8.2. The luster is metallic; the color, silver-white. The composition is $(Au,Ag)Te_2$, which yields 24.5 per cent gold and 13.4 per cent silver. Its reaction on charcoal is similar to that of calavarite, but the remaining globule is considerably lighter in color because of the presence of silver.

## PYRARGYRITE, $Ag_3SbS_3$; PROUSTITE, $Ag_3AsS_3$

Pyrargyrite and proustite are relatively rare but beautiful minerals known together as the *ruby-silver ores*. They are similar in crystal form and physical properties. The compositions are analogous in that the arsenic of proustite takes the place of the antimony of pyrargyrite.

PYRARGYRITE is rhombohedral and frequently shows well-developed faces of the prism and the rhombohedron. It also has a rhombohedral cleavage. The hardness is $2\frac{1}{2}$, and the specific gravity is 5.85. The luster is adamantine, almost metallic in some specimens, and the color deep red to black, giving it the name *dark ruby silver*. The streak is red.

PROUSTITE is quite like pyrargyrite in crystal form, cleavage, and hardness. Its specific gravity, 5.55, is slightly lower. The color, ruby red, brighter than pyrargyrite, gives it the name *light ruby silver*. The streak is red. Both minerals fuse easily on charcoal, pyrargyrite giving the dense white coating of antimony trioxide, and proustite giving the volatile sublimate of arsenious oxide with the garlic odor.

**Occurrence.** Of these two minerals, pyrargyrite is the more abundant and in some places is an important silver ore. Both are found in veins associated with other silver minerals at various places in Colorado, Nevada, and New Mexico. They were found with native silver at Cobalt, Ontario. Spectacular specimens have in the past come from Chanarcillo, Chile.

## TETRAHEDRITE $(Cu,Fe,Zn,Ag)_{12}Sb_4S_{13}$

**Habit.** Tetrahedrite is isometric and receives its name because the crystals are commonly tetrahedral in habit although often highly modified. Good crystals, as with many metallic minerals, are rare, and the mineralogist often has to content himself with massive pieces.

**Physical Properties.** Massive tetrahedrite can usually be recognized by the brilliant metallic luster and dark grayish black color and streak. It is frequently called *gray copper* by the miners. The hardness is $3-4\frac{1}{2}$; hence it is easily distinguished from magnetite, which is too hard to be scratched by the knife. The specific gravity varies from 4.6 to 5.1, depending on the proportion of the various elements present.

**Composition.** Tetrahedrite is a sulfantimonide of copper, iron, zinc, and silver, $(Cu,Fe,Zn,Ag)_{12}Sb_4S_{13}$. Copper is always the predominant metal. Although the formula seems complicated, it does not express the composition of the average specimen accurately, for usually some arsenic takes the place of some of the antimony. If arsenic takes the place of all the antimony, the mineral is called *tennantite*. The variety that is highly rich in silver is known as *freibergite*.

Tetrahedrite is easily fusible on charcoal and usually gives tests for both antimony and arsenic; if the remaining globule is touched

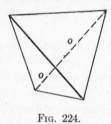
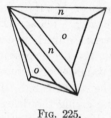

Fig. 224.                         Fig. 225.

Tetrahedrite.

with a drop of hydrochloric acid and heated again, it gives the azure-blue flame that indicates copper.

**Occurrence.** Tetrahedrite is widespread and is commonly associated with copper and silver minerals as well as pyrite, sphalerite, and galena. When it is high in silver, it becomes a valuable ore of that metal. In the United States it is mined at various places in Colorado, Montana, Nevada, Arizona, and Utah. Fine crystallized specimens have been produced in Bolivia.

## Enargite, $Cu_3AsS_4$

**Habit.** Enargite is orthorhombic, and crystals are usually prismatic with a striated prism zone. It is most commonly found in bladed or granular aggregates.

**Physical Properties.** A perfect prismatic cleavage is one of the outstanding features of enargite, for it is the only common mineral with two cleavage directions having metallic luster and black color and streak. The hardness is 3; the specific gravity is 4.43–4.45.

**Composition.** Copper sulfarsenide, $Cu_3AsS_4$. Easily fusible on charcoal, it gives the volatile white sublimate of arsenious oxide and characteristic garlic odor. If roasted on charcoal and moistened with a drop of hydrochloric acid and again heated, it gives the azure-blue copper flame.

**Occurrence.** Although enargite is a comparatively rare mineral, it is an important ore of copper at certain localities, where it is associated with other copper minerals. It is found in such an association at Butte, Montana, and at Bingham Canyon, Utah.

## OXIDES

In the oxide group are included several minerals of great economic importance, for the chief ores of iron, aluminum, tin, and chromium are oxides. In studying this group one should pay close attention to the physical properties, for by their use most of these minerals can be determined.

### Ice, $H_2O$

To many it may seem out of place to include ice here, but it is just as truly a mineral as diamond or quartz, even if it cannot be preserved in a mineral cabinet.

**Habit.** It occurs in crystalline forms with hexagonal symmetry, often of great complexity and beauty as in snow crystals. These, as stated on p. 15 are formed in the atmosphere directly from water vapor. The ice that makes the pellets of hail, not infrequently occurring with summer thunderstorms, are also occasionally in clusters of crystals, although usually they show a concentric concretionary structure. The ice of pools and ponds is crystalline, though usually the crystals are separately visible only in the first stages of the process of freezing. The process of crystallization begins, as everyone knows, when the temperature falls to 32° Fahrenheit (0° centigrade).

**Physical Properties.** The hardness of ice near the freezing point is $1\frac{1}{2}$, but it increases slightly at lower temperatures. The specific gravity is 0.92 and hence ice floats in water with a little more than nine-tenths of its bulk submerged. Water expands, therefore, on freezing and exerts a great force on confining surfaces. One consequence is the breaking of vessels, water pipes, etc., when the water they contain is frozen. In nature ice is for this reason a powerful agent in breaking up rock masses; the water creeps into the cracks, especially the narrow ones, and when it freezes the rock masses are slowly but surely wedged apart.

**Composition.** Water and ice consist chemically of hydrogen and oxygen, $H_2O$.

## Cuprite, $Cu_2O$

Cuprite is called *ruby copper* and *red copper ore* because of the fine red color of the crystals. It has served as an important ore of copper but is today only of minor importance compared to the copper sulfides.

**Habit.** Cuprite is isometric, and the crystals often show the cube, octahedron, and dodecahedron or combinations of these forms. (Figs. 226 to 228.) In one type of crystal the cubes are spun out into long threads, forming a matted mass of bright red

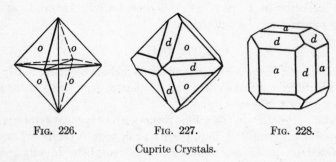

Fig. 226.     Fig. 227.     Fig. 228.

Cuprite Crystals.

hairs called *chalcotrichite* or *plush copper*. When they are examined closely with a glass, the threads are often seen to cross each other at right angles as if trying to build up skeleton cubes, the threads taking the direction of the cube edges. Common cuprite is massive, and the crystals are usually found in its cavities.

**Physical Properties.** The hardness of cuprite is 3½–4, and the specific gravity 6.0. The luster is adamantine but on some dark surfaces may look almost metallic. The color is red of various shades—ruby red in the clear transparent crystals—but the surface is often darkened and may appear nearly black. The streak is always brownish red.

**Composition.** Cuprous oxide, $Cu_2O$. A fragment heated on charcoal is easily robbed of its oxygen and reduced to metallic copper.

**Occurrence.** Cuprite is a mineral of secondary origin; that is, it has formed at or near the earth's surface by the oxidation of copper sulfide veins. It is associated with limonite and other secondary copper minerals, such as malachite, azurite, and native copper. In the United States cuprite has been found in fine crystals at Bisbee, Arizona. At some places cubes and octahedrons of malachite are found that are pseudomorphs after cuprite.

## ZINCITE, ZnO

Zincite is an important mineral at only one locality, Franklin, New Jersey, where it is associated with franklinite and willemite. These three are the zinc-bearing ore minerals at this famous locality, where they have been mined for over one hundred years and are still being mined on a large scale today.

**Habit.** Zincite is hexagonal, but crystals are rare. It is usually massive with a platy appearance.

**Physical Properties.** There is perfect prismatic cleavage; hence even a massive specimen shows flashing cleavage faces. The hardness is 4–4½, and the specific gravity 5.5. The color is deep red to orange-yellow, but the streak is always orange-yellow.

**Composition.** Zinc oxide, ZnO. The color of zincite is probably due to manganese, which is always present in small amounts, for pure ZnO is white.

## CORUNDUM, $Al_2O_3$

Corundum is a mineral that has attracted attention through the centuries, for the clear blue varieties make the *sapphire* and the clear red the highly prized *ruby*. Next to diamond it is the hardest mineral, and for this reason it has long been used as an abrasive. *Emery*, a natural mixture of fine-grained corundum and magnetite, was at first the only type of corundum used as an abrasive, but later coarsely crystalline corundum was mined and pulverized to be made into grinding wheels and similar tools.

The commercial importance of corundum is not so great today as it has been in the past, for most of the abrasive materials are now manufactured synthetically. The gem varieties also have been made in the factory and with such success that only the expert can tell the natural ruby and sapphire from the manufactured.

**Habit.** Corundum is hexagonal and when in distinct crystals it usually shows either the hexagonal prism or tapering pyramids (Fig. 229). It is frequently deeply striated as the result of repeated twinning on the rhombohedron.

**Physical Properties.** Corundum has no cleavage, but it may have as many as seven directions in which it breaks easily along parting planes. The most common of these are the three rhombohedral surfaces that intersect at almost right angles and yield fragments which appear cubic. The hardness is 9, and, therefore, it will scratch any other mineral except the diamond. The specific gravity is 4.0, which is high for a nonmetallic mineral and remarkably high for the oxide of a metal of such low density. The oxide of a metal is not often more dense than the metal itself; this density is obviously related to the high hardness. The luster, like that of most very hard minerals, is brilliant and adamantine in clear crystals; it may be dull in some massive varieties.

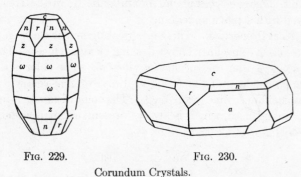

Fig. 229.                    Fig. 230.

Corundum Crystals.

The color is gray to brown in many of the common varieties, bright blue in the sapphire, red in the ruby. Gem corundum or other colors is frequently called by the names of other gemstones with the prefix *oriental*. For example, the purple is *oriental amethyst;* the yellow, *oriental topaz.* Some rare varieties of natural corundum when viewed in the direction of the *c* axis have a stellate opalescence and when cut into gems are the highly prized *star sapphire* and *star ruby.* Emery, as already stated, is a black natural mixture of corundum and magnetite, frequently so fine-grained as to appear homogeneous. It can be identified by powdering a small sample and removing the magnetite grains with a magnet, leaving behind the lighter-colored corundum.

**Occurrence.** Common corundum occurs in New Jersey, Pennsylvania, North Carolina, Georgia, and Montana. In Montana gem material also has been found both in river sands and in place. The world's most famous localities for rubies and sapphires are Burma, Siam, and Ceylon. Emery has long been mined in Greece and Turkey and during the last half of the nineteenth century was mined at Chester, Massachusetts.

## HEMATITE, $Fe_2O_3$

Hematite is named from the Greek word for blood, because many of its varieties are red and all give a red streak. Although alloys of magnesium and aluminum are for certain purposes replacing iron and steel, we are still living in the iron age, and iron is by far the most important metal in our present civilization. Several minerals are mined as ores, but hematite heads the list and thus may be considered commercially the most important of all the minerals. In the United States about 80 million tons of iron are produced each year, and nine-tenths of this comes from hematite.

FIG. 231.    FIG. 232.    FIG. 233.
Hematite Crystals.

It seems unnecessary to mention the uses of iron, for they are so numerous and encountered so frequently in our everyday life that they are familiar to all.

**Habit.** Hematite is rhombohedral, and in certain places beautifully formed crystals showing one or more rhombohedrons are found (Fig. 231). Other crystals are flattened parallel to the basal pinacoid and may appear as extremely thin flakes. In some specimens the plates are grouped in rosette forms (iron roses) as shown in Fig. 153, p. 58. When the flakes are arranged in a foliated micaceous mass, the aggregate is called *specular hematite* or *specularite*. Radiating aggregates, *kidney ore* (Fig. 160, p. 61), give reniform shapes. Hematite is usually earthy and shows no crystal form. The term *martite* is given to octahedral pseudomorphs of hematite after magnetite.

**Physical Properties.** The most outstanding characteristic of hematite in most specimens is its red color. However, crystals and the specular variety are black with a brilliant metallic luster, but the streak of these, as of the massive varieties, is red. The earthy kind, dull in luster, is the *red ocher* used for making paint.

The hardness of the crystals is about 6, and they are hence too hard to be scratched by a knife.  One should be cautioned in taking the hardness of the specular varieties, for in drawing the knife blade across the specimen it is easy to separate the tiny flakes and this separation may be mistaken for a scratch.  In the red earthy varieties the hardness may be as low as 1.  The specific gravity is 5.26 for the crystals.

Crystallographically hematite is similar to corundum and like it may show, in coarsely crystallized specimens, a rhombohedral parting.  The intersections of these three parting directions are nearly at right angles, and fragments resemble minerals with cubic cleavage.

**Composition.**  Hematite is ferric oxide, $Fe_2O_3$, and if pure contains 70 per cent iron.  When heated in the reducing flame it becomes strongly magnetic.

**Occurrence.**  Hematite is a widespread mineral occurring in scattered grains through many of the igneous rocks.  It is one of nature's most abundant pigments, giving the red color to sandstones such as those in the Grand Canyon.  Large masses of specular ore are associated with metamorphic rocks.  Oölitic hematite is found in extensive beds of sedimentary origin.  The chief deposits in the United States, which supply a high percentage of the world's iron ore, are grouped around the northwestern and southern shores of Lake Superior in Minnesota, Wisconsin, and Michigan.  It is mined in Alabama also, near Birmingham.  The most beautiful crystals have come from the island of Elba and from Switzerland.

## ILMENITE,  $FeTiO_3$

Ilmenite, also called *titanic iron ore*, is related to hematite, but titanium takes the place of half of the iron of hematite.  Tremendous masses of ilmenite that are potential iron ores are known, but, because of difficulties in smelting, it is not extensively used for that purpose.  It is, however, a source of titanium, and titanium oxide is being used more and more as a paint pigment.

**Habit.**  Ilmenite is rhombohedral with the crystals, usually tabular, showing prominent basal planes.  It is most commonly massive.  In some beach sand ilmenite is abundant as the black grains.

**Physical Properties.**  The hardness is $5\frac{1}{2}$–6, and the specific gravity, 4.7.  The luster is metallic; the color, iron-black; the streak, black to brownish red. It can be distinguished from hematite by the streak and from magnetite by its lack of magnetism. Some specimens however, may be slightly magnetic without heating.

**Composition.** Ferrous titanate, $FeTiO_3$. It becomes strongly magnetic after heating.

**Occurrence.** Ilmenite is widely distributed as an accessory mineral in igneous rocks, but it also occurs in large masses associated with metamorphic rocks or as a magmatic segregation. In the United States large bodies of it are found associated with magnetite in the Adirondack region of New York. Commercial deposits are found also in Labrador, India, and Norway.

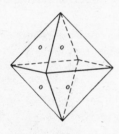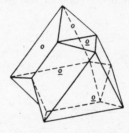

Fig. 234.                         Fig. 235.

Spinel Crystals.

## SPINEL, $MgAl_2O_4$

Spinel is a rather rare mineral that in places is found in beautifully colored transparent crystals that are used as gems.

**Habit.** Spinel is isometric and is characteristically found in octahedral crystals. Twinned octahedrons (Fig. 235) are common, and thus the name *spinel twin* is given to this type.

**Physical Properties.** The hardness is 8, or as great as that of topaz, and the specific gravity is 3.5 to 4.1. The color is not diagnostic for it may be white, red, blue, green, lavender, brown, or black. The red variety is known as *spinel ruby* or *balas ruby* and should not be confused with the true ruby, corundum. It is nonmetallic with a vitreous luster and usually translucent, although the dark varieties appear almost opaque. The streak is white.

**Composition.** The typical composition is $MgAl_2O_4$, but iron, manganese, and chromium may be present in varying amounts.

*Gahnite* is green zinc spinel in which zinc has taken the place of all the magnesium.

**Occurrence.** Spinel is a metamorphic mineral and thus is found in crystalline limestones and schists. Because of its chemical and mechanical stability, it is frequently found as pebbles in stream sands. Gem spinel is found associated with gem corundum in Ceylon, Siam, and Burma.

## MAGNETITE, $Fe_3O_4$

Magnetite is an important ore of iron but ranks considerably behind hematite in importance. Its name suggests its most striking characteristic, its magnetism. All kinds are strongly attracted by a magnet, and one variety, *lodestone*, is a powerful magnet itself.

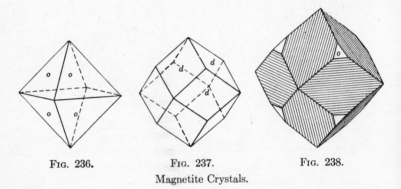

FIG. 236.          FIG. 237.          FIG. 238.

Magnetite Crystals.

It has north and south poles and the power of picking up particles of iron or steel (Fig. 180, p. 89). When suspended it aligns itself with its poles, north and south, like a compass needle. The magnetism of lodestone excited the imagination of the early poets, and they attributed tremendous powers to it, great enough to pull nails out of ships!

**Habit.** Magnetite is isometric, and crystals usually are octahedral, more rarely dodecahedral. It is most commonly simply massive as a granular aggregate.

**Physical Properties.** Although magnetite has no cleavage it shows on some specimens an octahedral parting that will yield octahedral fragments. The hardness is 6, and the specific gravity 5.18; both properties are close to those of hematite. The luster is metallic, usually very brilliant, and the color iron-black. The streak also is black, an important characteristic, for it distinguishes magnetite at once from hematite, which, though at times iron-black, has a *red* streak. As previously stated, its magnetism is its most important physical property.

**Composition.** The formula for magnetite is $Fe_3O_4$ which yields 72.4 per cent iron, slightly more than hematite.

**Occurrence.** Like hematite and ilmenite, magnetite is found scattered through igneous rocks as an accessory mineral. In fact, an effort was made to extract a few per cent of it from a rock in New Jersey for use as an iron ore. Large masses of it have been found in eastern United States, and, until the discovery of the Lake Superior ores about the middle of the nineteenth century, it was the chief iron ore. Important commercial deposits are today mined in northern Sweden and in Norway. Lodestone is found in crystals at Magnet Cove, Arkansas.

## FRANKLINITE, $(Fe,Zn,Mn)(Fe,Mn)_2O_4$

Franklinite, so called from its only important locality, Franklin, New Jersey, is a mineral much resembling magnetite in form, color, and general appearance. It is, however, only feebly magnetic, if at all, and has a brown, not a black, streak. The hardness is 6, and the specific gravity 5.15. It contains besides iron both zinc and manganese, and hence is valuable as a zinc ore and for making spiegeleisen, an alloy of iron and manganese employed in the making of steel. Franklinite can usually be identified by its characteristic association with willemite and zincite. (See p. 163.)

## CHROMITE, $FeCr_2O_4$

Chromite, or *chromic iron ore*, is the only important source of chromium, an element that in recent years has been in great demand. Its chief use is in making chrome steel, an alloy not only hard and tough but also resistant to chemical attack. It is used also in making stainless steel, resistance wire in electrical equipment, and high-speed cutting tools. The most familiar use is as a material for plating. Automobile accessories, plumbing fixtures, and hardware after being plated with chromium can be polished to a brilliant surface. Because of its refractory nature chromite is crushed and then pressed into bricks for lining open-hearth furnaces.

FIG. 239.   Franklinite Crystals in Calcite, Franklin, New Jersey.

**Habit.**   Chromite is isometric and when in crystals is octahedral resembling magnetite, but crystals are rare, and it is usually massive.

**Physical Properties.**   The hardness is $5\frac{1}{2}$, and the specific gravity 4.6.   The color is iron-black to brownish black; the streak, dark brown.   The luster of some specimens is metallic, of others, submetallic with a pitchy appearance—an important characteristic for the sight determination of chromite.

**Composition.**   The formula is $FeCr_2O_4$, but chromite may contain some magnesium, aluminum, and iron.   A small grain of chromite in a borax bead imparts a green color to the bead in both the oxidizing and reducing flames.

**Occurrence.**   Chromite is usually associated with peridotite rocks, or with serpentines, and in this way has been found in Pennsylvania, Maryland, and California.   It has been found more recently in the Stillwater Igneous Complex in Montana.   Very little is mined in the United States, and most of the world's production comes from Southern Rhodesia, Greece, India, and the U.S.S.R.

## CHRYSOBERYL, BeAl₂O₄

Chrysoberyl is a rare mineral occurring in pegmatites and in mica schists. Its only use is as a gem and as such is found in Brazil, Ceylon, and the Ural Mountains. The variety known as *alexandrite* is of interest because it changes from emerald-green in day light to red in artificial light. *Cat's-eye* or *cymophane* is a variety that, when cut into a gemstone, shows a narrow beam of light that changes position with movement of the stone.

**Habit.** Chrysoberyl is orthorhombic and crystals are usually flattened parallel to the front pinacoid. Other crystals are twinned giving them a hexagonal appearance (Fig. 240).

**Physical Properties.** Chrysoberyl has prismatic cleavage and thus has two directions of easy breaking. The hardness is 8½ and hence chrysoberyl will scratch topaz. The specific gravity is 3.65–3.8. The common type has a greenish yellow color, slightly resembling beryl, whence it takes its name, for chrysoberyl means *golden beryl*.

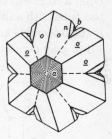

FIG. 240. Chrysoberyl Twin.

**Composition.** Beryllium aluminate, BeAl₂O₄. It is an infusible and insoluble mineral, and one must rely mostly on physical tests for identification.

## CASSITERITE, SnO₂

Cassiterite, or *tinstone*, is almost the sole source of tin. The only other ore mineral of tin is the rare sulfide of tin, copper, and iron called *stannite*. Tin is in great demand, and the world's supply appears to be limited. Its chief use is in the manufacture of *tin plate* to be made into cans for food containers. It is used also with lead in solder, with antimony and copper in babbitt metal, and with copper in bronze.

**Habit.** Cassiterite is tetragonal and commonly forms in prisms and pyramids (Fig. 241). It is found frequently in twins (Fig. 242) but is usually massive. Some cassiterite shows a reniform shape with radiating fibrous appearance and is called *wood tin*.

**Physical Properties.** Cassiterite is remarkable for its hardness, 6–7, and still more for its high specific gravity, about 7. This is unusually high for a nonmetallic mineral. The luster is adamantine to submetallic; the color is usually brown to black, more rarely yellow or white. The streak is white.

**Composition.** Tin dioxide, $SnO_2$. Pure tin dioxide is white; it is the presence of small amounts of iron that makes the mineral dark. Cassiterite can usually be recognized by its physical properties, but, if one is uncertain, the following test can be used. Drop a fragment of the mineral and a piece of metallic zinc into a test tube with a little dilute hydrochloric acid and warm it gently. If the mineral is cassiterite, the specimen will become coated with a dull gray deposit of metallic tin.

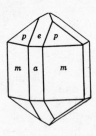

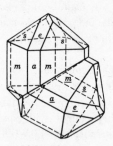

Fig. 241.                    Fig. 242.

Cassiterite Crystals.

**Occurrence.** Cassiterite usually is associated with granitic rocks, where it may be in veins or scattered through the rock, often in inconspicuous particles. These tiny grains can be separated by the same process that nature has used in making stream tin, that is, after the rock has been crushed, the lighter material can be washed away leaving behind the heavy cassiterite. Indeed most of the world's tin is obtained by washing sand and gravel that contain cassiterite, known as *stream tin*, washed in from a near-by primary source. The deposits in the Malay States, Netherlands East Indies, and Siam, from which most of our tin comes, are of this type. Tin is mined also in Bolivia, and formerly Cornwall, England, produced large amounts. In the United States cassiterite has been found in numerous places, notably in the Black Hills, South Dakota, but at none of these is it in sufficient quantity to be mined.

## Rutile, $TiO_2$

$TiO_2$ exists in three polymorphic forms, of which rutile is the most common. The others are *anatase* or *octahedrite*, which is tetragonal, and *brookite*, which is orthorhombic. Paramorphs of one after the other are common, and it is difficult for the beginner to determine which mineral he is examining.

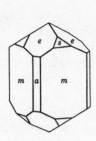

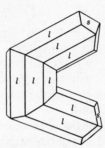

Fig. 243.

Fig. 244.

Rutile Crystals.

Rutile together with ilmenite and sphene is a source of titanium for paint pigment, for electrodes in arc lights, and for coloring porcelain and false teeth. Rutile alone is used for coating welding rods for welding steel. Synthetic rutile is nearly colorless and can be fashioned into very attractive gemstones.

**Habit.** Rutile is tetragonal and is commonly found in crystals, the forms of which are shown in Fig. 243. It is frequently in elbow twins (Fig. 244). Crystals are sometimes slender, and a network of them may penetrate quartz crystals to form specimens of great beauty.

**Physical Properties.** The hardness is 6–6½, and the specific gravity 4.2. The color varies from reddish brown to red; some specimens may be nearly black, but even they let a little reddish light through a thin splinter. The luster is adamantine to sub-metallic; the streak, pale brown.

**Composition.** Titanium dioxide, $TiO_2$. A small amount of iron may be present.

**Occurrence.** Rutile is a common though minor constituent of many rocks, such as granite, granite pegmatite, gneiss, and schist. It is both mechanically stable and chemically inert; hence, on the disintegration of rocks, it may be washed many miles, where it accumulates as a constituent of black sands associated with magnetite, zircon, and monazite.   Disseminated rutile has been extracted from rocks at Magnet Cove, Arkansas, and Amherst and Nelson County, Virginia.   Large fine crystals have come from Graves Mountain, Georgia.

## PYROLUSITE, $MnO_2$

Pyrolusite is an important mineral for it is the chief ore of manganese, a vital element in the production of steel.   It is used with iron to make spiegeleisen which, when introduced into the steel batch, removes oxygen thus preventing oxidation of the steel. Ninety per cent of the manganese produced is used in the manufacture of steel.

Because of the large amount of oxygen in pyrolusite, it is sometimes used in the laboratory as a source of that gas.   The glassmaker also employs it to remove the unwanted color from glass, and for this reason it takes its name from two Greek words meaning *fire* and *to wash*.   It has many other uses, such as an oxidizer in the manufacture of chlorine and bromine, a drier in paints, and in electric dry cells and batteries.

**Habit.**   Pyrolusite is tetragonal, but it is rarely in crystals; when they are found they go under the name of *polianite*.   Pseudomorphs of pyrolusite after manganite are common.   It is usually massive or in radiating fibers and often forms dendrites (Fig. 164, p. 63) along cracks in rocks.

**Physical Properties.**   The hardness is 1–2, soft enough to soil the fingers;   the specific gravity is 4.75.   The luster is metallic; the color and streak, iron-black.   One can distinguish it from the other manganese oxides by the streak, for that of manganite is brown, and that of psilomelane is brownish black.

**Composition.**   Manganese dioxide, $MnO_2$, and, although it may contain a little water, the amount is small compared with that in manganite and psilomelane.   When pyrolusite is powdered and dissolved in a bead of sodium carbonate, the bead becomes bluish green, indicating manganese.

**Occurrence.**   Pyrolusite, like the other manganese oxides, is a secondary mineral formed by the oxidation of manganese-bearing silicates and carbonates.   Manganese ores have been mined in Virginia, Georgia, Arkansas, and Tennessee; but most of the requirements of the United States are met by imports. The chief producing countries are Russia, South Africa, Brazil, India, and Cuba.

## DIASPORE, AlO(OH)

Diaspore is a rare mineral formed usually as a decomposition product of corundum and associated with that mineral, as at Chester, Massachusetts. It is one of the constituents of bauxite, but it is so finely divided that it is impossible for the beginner to recognize it.

**Habit.** Diaspore is orthorhombic and is usually in thin crystals flattened parallel to the side pinacoid. It is also massive.

**Physical Properties.** There is perfect cleavage parallel to the side pinacoid. The hardness is $6\frac{1}{2}$-7; the specific gravity, about 3.4. The color may be white, gray, yellowish, or greenish; the luster is vitreous, but pearly on the cleavage face.

**Composition.** The formula for diaspore is AlO(OH). It is infusible and insoluble, and is usually recognized by its cleavage, platy habit, and its hardness.

## GOETHITE, FeO(OH); LIMONITE, FeO(OH)·$n$H$_2$O

Goethite and limonite can be dealt with together, for most of their properties are similar. Until rather recent years goethite has been considered the minor of the two minerals, and limonite has been said to occur in great abundance. Today, however, as the result of new methods of study, the situation is reversed; goethite is the common mineral, and limonite is relatively rare. The distinction between the two minerals is made almost entirely on the basis of crystal structure. If the substance is crystalline, even if an x-ray study is necessary to determine it, it is goethite. Only if it is amorphous, showing no crystal structure at all, it is called limonite. The name, limonite, however, is frequently used as a field term for any brown iron oxide the identity of which is not readily apparent.

At certain localities goethite is an important iron ore. It is the principal constituent of the iron ores of Alsace-Lorraine. In the Mayari and Moa districts of Cuba minable deposits of goethite have been formed by the alteration of iron-rich rocks.

**Habit.** Goethite is orthorhombic. Crystals are rare but when found are usually prismatic with vertical striations. It is usually massive, reniform, or stalactic and commonly has a radiating habit (Fig. 245). Limonite is amorphous with no crystal structure.

**Physical Properties.** There is perfect cleavage parallel to the side pinacoid. In the fibrous aggregates the cleavage can be seen parallel to the length of the fibers. The hardness is 5–5½; the

Fig. 245. Radiating Goethite, Negaunee, Michigan.

specific gravity, 4.37. Goethite is frequently impure with the specific gravity as low as 3.3. The luster is adamantine to dull. The color is yellowish brown to dark brown; the streak, yellowish brown. Both goethite and limonite are characterized by a yellowish brown streak.

**Composition.** $FeO(OH)$, which gives 62.9 per cent of iron, a smaller percentage than in hematite and magnetite. Limonite is of similar composition but has additional water in indefinite amounts; it may contain also manganese oxides and clay. Both minerals become strongly magnetic after heating.

**Occurrence.** Goethite forms under a wide variety of conditions, and is usually present where iron minerals have been exposed to oxidizing agents. Thus it forms the iron-rich surface over ore bodies containing pyrite; the iron ores of Cuba are largely goethite formed by the alteration of serpentine.

The word limonite comes from the Greek meaning *meadow* because it is often found in marshy places; in fact one type is called *bog iron ore*. A careful study of such material would probably

Fig. 246. Manganite Crystals, Ilfeld, Germany.

show that it is mostly goethite. Deposits of bog iron ore were mined during the early days in the United States in Massachusetts, New York, Pennsylvania, and Virginia but were low-grade because of the clay and other impurities present. When pyrite is exposed to atmospheric conditions, it tends to alter to limonite and thus one of the commonest types of pseudomorphs is limonite after pyrite.

## Manganite, MnO(OH)

Manganite, like pyrolusite, is a manganese oxide formed by secondary processes. The two minerals are thus found together and with psilomelane. Manganite can be considered a minor ore of manganese.

**Habit.** Manganite is orthorhombic, and crystals of it are common in brilliant prisms (Fig. 246) and in fibrous, radiated masses. One should not be hasty about identifying manganite by crystal form alone, for it commonly alters to pyrolusite.

**Physical Properties.** There is perfect cleavage parallel to the side pinacoid. The hardness is 4; the specific gravity, 4.3. The color is steel-gray to iron-black; the streak, dark brown. It is by means of the streak that one is able to distinguish manganite from pyrolusite, which gives a black streak.

**Composition.** Manganite is a basic manganese oxide, $MnO(OH)$. If manganite is powdered and dissolved in a sodium carbonate bead, the fusion is bluish green, indicating manganese.

**Occurrence.** As noted above, manganite is associated with the other manganese oxides, all of which are of secondary origin; that is, they have formed at the expense of earlier minerals such as manganese carbonate or silicate. Manganite frequently is found altered to pyrolusite and hence one must test the hardness and the streak rather than rely on crystal form alone for identification. In the United States it is found at Negaunee, Michigan.

## BAUXITE

For many years mineralogists listed bauxite as a mineral species and gave it a definite chemical composition. However, in recent years it has been learned that bauxite is a mixture of several minerals and is, therefore, more of a rock name than a mineral name. It is included here among the minerals because as the ore of aluminum it is an important commercial substance.

The uses of aluminum are so many and so common that it is hardly worth noting them. For purposes for which it is necessary to have strength in a light-weight material aluminum has found many uses. It is used in the construction of airplanes and airplane engines, in automobiles, and in railroad cars. All are familiar with the cooking utensils, other household appliances, and furniture made from aluminum. It is a good electrical conductor and each year is replacing copper more and more for transmission lines. Other uses are in aluminum foil and in paints.

**Habit.** As previously mentioned, bauxite is not a mineral and therefore has no crystal system. It is usually found in pisolitic aggregates and in round concretionary grains (Fig. 247); it occurs also in claylike masses.

**Physical Properties.** The hardness varies from 1 to 3; the specific gravity, from 2 to 2.5. The luster is usually earthy; the color, white, gray, yellow, or red.

**Composition.** Since bauxite is a mixture of several minerals it is impossible to give a formula for it, but some of it approaches closely the composition of the mineral gibbsite, $Al(OH)_3$. The

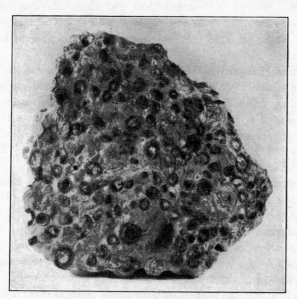

FIG. 247. Pisolitic Bauxite, Bauxite, Arkansas.

considerable water present can be driven off by heating in a closed tube. If it is heated on charcoal and then touched with a drop of cobalt nitrate and heated again, it assumes a blue color, the test for aluminum in infusible substances.

**Occurrence.** Bauxite is a substance of secondary origin formed by the disintegration of rocks under tropical or subtropical conditions. In this process the silica is carried away and the hydrous aluminum oxides are left behind. If iron is present, it also will remain behind, usually as limonite, which, if present in large amounts, renders the material unusable as an ore of aluminum. In the United States bauxite is mined in Arkansas, Alabama, and Georgia, but more than 50 per cent of the bauxite used is imported, mostly from Dutch Guiana and British Guiana. The name bauxite is taken from the important district at Baux, France.

## BRUCITE, $Mg(OH)_2$

Brucite until recent years has been of interest to the mineralogist only and has had no particular commercial significance. However, it is at present mined and heated to produce MgO for use as refractory material.

**Habit.** Brucite is rhombohedral, but rhombohedrons are not prominent crystal forms, and crystals are usually tabular parallel to the basal pinacoid. It is commonly in foliated or massive aggregates.

**Physical Properties.** There is perfect basal cleavage which resembles somewhat that of mica, but the folia are not elastic and, when bent, will not return to their initial position as in mica. The hardness is $2\frac{1}{2}$; the specific gravity, 2.4. The luster on the base or cleavage face is pearly. Elsewhere or on a massive specimen it is waxy. The color is white, pale gray, or light green. Brucite is sectile.

**Composition.** Magnesium hydroxide, $Mg(OH)_2$. Brucite is infusible but when heated in the blowpipe flame gives off water, making the fragment chalky white.

**Occurrence.** Brucite is a secondary mineral formed by the decomposition of magnesium-bearing rocks, usually serpentines. It is thus associated with other minerals, such as hydromagnesite and magnesite, which are formed by the same process. In the United States it has been known for many years at the Tilly Foster mine at Brewster, New York, and recently it has been mined near Luning, Nevada.

## PSILOMELANE

Psilomelane is another secondary manganese oxide associated with pyrolusite and manganite and is next in importance to pyrolusite as an ore of manganese. (See pyrolusite, p. 174.)

**Habit.** Psilomelane is orthorhombic, but crystals are extremely rare and very small. It usually is found massive or in botryoidal or stalactitic aggregates.

**Physical Properties.** The hardness is 5–6, much greater than that of pyrolusite or manganite; the specific gravity is 3.7–4.7. The color is black but the streak is brownish black, thus distinguishing it from the other manganese oxides.

**Composition.** Psilomelane is essentially a hydrous manganese oxide, but it contains many other elements, particularly barium. Varying amounts of magnesium, calcium, nickel, cobalt, and copper may be present. When a small fragment is fused with sodium carbonate, it gives a bluish green bead, the test for manganese.

**Occurrence.** Psilomelane is found associated with pyrolusite and has a similar origin.

# HALIDES

The halides are compounds of chlorine, fluorine, bromine, and iodine. Many of them are very rare, and only three chlorides and two fluorides are described here.

## HALITE, NaCl

Halite is familiar to us all as the common salt used in cooking and on the table. Since it is essential to the life of man, it has been sought, traded, and fought for throughout the history of man. Besides its culinary use, salt is a preservative and furnishes sodium to the chemical industry for the manufacture of sodium carbonate (soda ash). This compound is used in making soap and glass as well as in bleaching, dyeing, and refining of oil.

**Habit.** Halite is isometric and crystallizes in fine clear cubic crystals. It is most abundantly found in granular, cleavable masses known as *rock salt*. Sometimes the crystals have the skeleton or hopper-shape illustrated in Fig. 248.

**Physical Properties.** There is perfect cubic cleavage, which is seen in the aggregates as well as in the crystals. The hardness is $2\frac{1}{2}$; the specific gravity, 2.16. It has a vitreous luster and is colorless when perfectly pure. Impurities may give it various shades of red and yellow; occasional patches of a fine deep blue are seen in the clear crystals.

**Composition.** Halite is sodium chloride, NaCl. The deep yellow color that it gives to the blowpipe flame is the characteristic test for sodium. Halite is one of the few important minerals that are readily soluble in water and hence give a decided taste.

**Occurrence.** Salt is commonly found in beds as a sedimentary rock, where it has formed by the evaporation of sea water; thus it is found with other minerals such as gypsum and sylvite that form in the same way. The ocean is hence the important source as well as the great storehouse of the salt of commerce. Sodium chloride is present, and in even greater concentration, in some inland seas also, such as the Great Salt Lake in Utah, the Dead and Caspian seas, and many others. In some places salt is actually mined, as in Austria and Louisiana; in other locations wells are drilled to the salt bed from which brine is pumped up and evaporated to yield

Fig. 248. Halite.

the salt. Salt is produced from such wells in New York, Michigan, and Ohio. In Texas and Louisiana many vertical pipelike bodies of salt, which have apparently punched their way upward from an underlying bed of salt, have been discovered. These *salt domes* are searched for because of the petroleum frequently associated with them.

## Sylvite, KCl

Sylvite is a salt similar to halite in its crystal form, its occurrence, origin, and mineral association, but it can be distinguished from halite by its more bitter taste. It is a very important mineral for it is the chief source of potassium compounds, which are used extensively as fertilizers.

**Habit.** Sylvite is isometric, and crystals frequently show the cube and octahedron in combination. It is most commonly found in granular, cleavable masses.

**Physical Properties.** There is perfect cubic cleavage. The hardness is 2.0; the specific gravity, 1.99. It is colorless or white when pure but may be colored various shades of blue, yellow, or red by impurities. It is more soluble in water than halite, and can be distinguished by its bitter taste.

**Composition.** Sylvite is potassium chloride, KCl. It fuses easily before the blowpipe, imparting a violet color to the flame due to potassium. If sodium is present, the violet color will be seen only if the yellow sodium flame is filtered out by means of a piece of blue glass.

**Occurrence.** Sylvite, like halite, is formed in beds resulting from the evaporation of sea water. Inasmuch as sylvite is more soluble than halite, it is precipitated later; if evaporation does not continue to almost complete dryness, no sylvite may be precipitated. It is, therefore, a much less common mineral than halite. Sylvite has been mined in large amounts at Stassfurt, Germany, and in recent years large deposits have been worked near Carlsbad, New Mexico.

## CERARGYRITE, AgCl

Cerargyrite, translated "from Greek" into English, means *horn silver*, a name frequently applied to it because of its hornlike appearance and the ease with which it is cut by a knife. In certain localities it has been an important ore of silver.

**Habit.** Cerargyrite is isometric, but crystals are rare. It is usually found in scales, plates, or masses resembling wax.

**Physical Properties.** The hardness is 2–3; the specific gravity, 5.5. It is remarkable for being perfectly sectile; that is, it can be cut with a knife like a piece of lead or wax. The luster is adamantine; the white or pale gray color rapidly darkens to violet-brown on exposure to light.

**Composition.** Cerargyrite is silver chloride, AgCl. Before the blowpipe on charcoal it fuses very easily, yielding a globule of silver. Other minerals closely related to cerargyrite and having similar physical properties are: *embolite*, $Ag(Cl,Br)$; *bromyrite*, AgBr; and *iodyrite*, AgI.

**Occurrence.** Cerargyrite is a secondary mineral and is found in the upper, near-surface portions of silver veins. It was an important ore mined at Leadville, Colorado, and at the Comstock Lode, Nevada.

## CRYOLITE, $Na_3AlF_6$

Cryolite takes its name from two Greek words that mean *ice stone*, because blocks of the mineral often have the appearance of slightly clouded blocks of ice.   It is easily fusible and, when in the molten state, is an excellent solvent.   It is for this reason used to clean metal surfaces and to dissolve the aluminum oxides used in the electrolytic process for the production of aluminum.   During the latter part of the nineteenth century when metallic aluminum was a rarity, cryolite was the ore of that metal.

FIG. 249.   Cryolite.

**Habit.**   Cryolite is monoclinic, although crystals have nearly cubic angles (Fig. 249).   It is most commonly found in massive form.

**Physical Properties.**   Cryolite has no cleavage but three directions of parting, which, unless examined carefully, may be confused with cubic cleavage.   The hardness is $2\frac{1}{2}$; the specific gravity, 2.95–3.0.   The luster is vitreous to greasy; the color, snow-white.

**Composition.**   Cryolite is sodium aluminum fluoride, $Na_3AlF_6$. It fuses easily in the candle flame, coloring the flame an intense yellow, the test for sodium.

**Occurrence.**   Cryolite is a rare mineral, and the only locality where it occurs in quantity is at Ivigtut in southwestern Greenland. Here it is characteristically associated with siderite, galena, and chalcopyrite.   It has also been found in minor amounts at the foot of Pikes Peak in Colorado.

## FLUORITE, $CaF_2$

Fluorite or *fluor spar* is one of the most beautiful minerals, occurring in well-formed crystals of many different colors.   It is sometimes cut into gems, but, because of its rather low hardness, the stones are not durable.   The name comes from the Latin word meaning *to flow*, since it melts more easily than other gemstones which it resembles.

The chief use of fluorite is as a flux in making steel, but it is also used in making opalescent glass, in enameling cooking utensils, and in the preparation of hydrofluoric acid.   Small amounts of fluorite of high purity are used in making prisms and lenses for optical equipment.

**Habit.**   Fluorite is isometric, and cubic crystals and groups of crystals are common (Fig. 250).   It is found in penetration twins also, such as those shown in Fig. 253, in which the angles of one cube project from the faces of another.   The position of the

FIG. 250. Fluorite, Cumberland, England.

crystals is such that, if one of them were revolved 180° about a line joining opposite angles, they would be brought into parallel position. On the edges of the cubes pairs of narrow faces of the tetrahexahedron may be seen on some crystals (Fig. 251). Less commonly six small faces of the hexoctahedron are seen on the solid angles (Fig. 252).

Octahedral crystals are also found but when they are examined closely it can be seen that most of them are built up of minute cubes. The cubic habit, however, is so characteristic that, when this form is encountered in a nonmetallic mineral, fluorite is at once suggested to the careful mineralogist. Fluorite occurs in fibrous or columnar aggregates also and, in one variety with the colors arranged in bands, known as *blue john*, is used as an ornamental stone. There are granular and closely compact varieties also.

**Physical Properties.** There is perfect octahedral cleavage, so that fine crystals must be handled carefully to prevent breaking off the corners of the cubes. The hardness of fluorite is 4; its specific gravity, 3.2. The great variety in color, embracing many shades of green, purple, yellow, and red, has already been mentioned.   It may also be colorless, dark brown, and black.   Crystals are usually transparent, and some show a banding due to the deposition of successive layers of differently colored material.

Some fluorite shows a beautiful fluorescence when exposed to ultra-violet light.   The phenomenon of fluorescence was first observed in fluorite and takes its name from the mineral.   Some fluorite exhibits *thermoluminescence* also, the property of emitting visible light when heated.   The variety *chlorophane* is so named because of the beautiful green light that it emits when heated. In some varieties the blow of a hammer is enough to make a mass yield a faint but beautiful light (triboluminescence) for hours afterward.

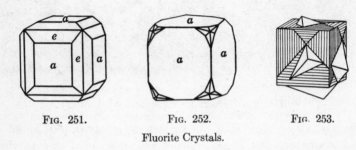

Fig. 251.              Fig. 252.            Fig. 253.

Fluorite Crystals.

**Composition.**   Fluorite is calcium fluoride, $CaF_2$.   When powdered and warmed with sulfuric acid, it gives off hydrofluoric acid, which will etch the test tube containing it.   It is an interesting experiment to cover a watch glass with a layer of wax, then with a fine point make a design by removing part of the wax.   If fluorite sulfuric acid mixture is placed on the wax, the design will be etched into the glass.

Fluorite usually flies to pieces violently when heated before the blowpipe, but, when pulverized, it can be fused easily.

**Occurrence.**   Fluorite is a widespread mineral and is commonly found with metallic ore minerals.   It is then said to form the *gangue* of the ore.   Beautiful specimens have been found in Derbyshire and Cumberland, England, and in the Freiberg mining district of Saxony.   The most important deposit in the United States is in southern Illinois at Rosiclare and Cave-in-Rock. Smaller deposits are in Colorado, New Mexico, and New York.

# CARBONATES

The carbonates are an interesting group because of the similarities shown by many of the minerals, thus illustrating well the reason for describing mineral species according to a chemical classification. With the exception of the copper carbonates, malachite and azurite, all the common carbonates fall into two groups: (1) the *calcite group;* (2) the *aragonite group.* The minerals of each group are called *isostructural* and show a close relationship to each other that cannot be brought out well if they are scattered through the book and listed under the metal dominant in each.

## CALCITE, CaCO₃

Calcite gives its name to the group of rhombohedral carbonates, which includes dolomite, magnesite, siderite, rhodochrosite, and smithsonite. All these minerals have similar crystal forms and perfect rhombohedral cleavage; the angles between the cleavage faces of the different species vary from 72° to 75°.

Calcite has more varieties that occur in abundance than any other mineral except quartz. Each of them is described under the heading *Varieties.*

**Habit.** Calcite is rhombohedral and crystallizes in a great variety and complexity of forms. The fundamental rhombohedron (Fig. 255), the flat rhombohedron (Fig. 254), and the scalenohedron (Fig. 263) are common forms. There are other rhombohedrons lengthened in the vertical direction as shown in Figs. 256 and 257. Figure 258 represents the hexagonal prism; Fig. 261 is the same with the obtuse rhombohedron *e* of Fig. 254 in combination. Figure 264 is a scalenohedron twinned on the base. Crystals forming in scalenohedrons or acute rhombohedrons are often called *dogtooth spar.* Besides the crystals illustrated, there are many other combinations of faces, some highly complicated, that can be deciphered only by one who has a thorough knowledge of crystallography. The types of calcite that do not show crystal forms are mentioned under *Varieties.*

**Physical Properties.** The outstanding characteristic of calcite is its perfect rhombohedral cleavage with an angle of 75° between the faces. Whatever the crystal form, a mass, whether large or

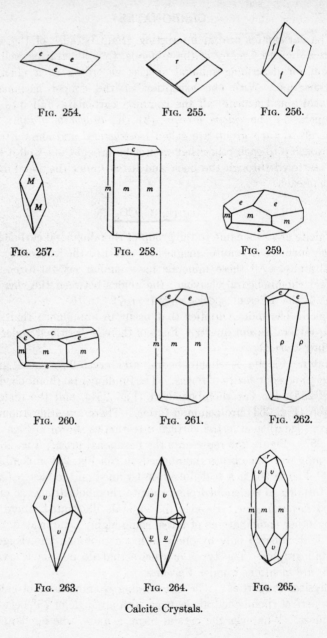

FIG. 254.        FIG. 255.        FIG. 256.

FIG. 257.        FIG. 258.        FIG. 259.

FIG. 260.        FIG. 261.        FIG. 262.

FIG. 263.        FIG. 264.        FIG. 265.

Calcite Crystals.

small, breaks easily under the blow of a hammer into fragments, all of which show the angles of Fig. 255. Calcite is frequently twinned, and crystals may show a parting parallel to the twinning lamellae that is at the angle of the negative rhombohedron (Fig. 259). The remarkable experiment by which twinning may be imparted to a cleavage fragment is mentioned on p. 56.

Calcite on most of its crystal faces and on the cleavage face has a hardness of 3, but on the basal pinacoid *c* (Fig. 258) it has a

*Ward's Natural Science Establishment*

Fig. 266. Calcite Crystal, Joplin, Missouri.

hardness of 2 and can be scratched by the finger nail. The specific gravity is 2.72. The luster is usually vitreous, but some varieties are dull to earthy. It is usually white or colorless but may be gray, red, yellow, green, or blue. Certain impurities may make it brown or black. One can see, therefore, that color cannot be used as a criterion for identification.

The clear colorless calcite, such as that originally brought from Iceland, is called *Iceland spar*, and is used for optical prisms because of its remarkable double refraction, or power of dividing a ray of light passing through it into two separate rays, so that a line seen through it appears double.   This phenomenon has been described and illustrated on p. 88.

**Composition.**   Calcite is calcium carbonate, $CaCO_3$.   One of the best tests for calcite is to touch it with a drop of hydrochloric acid and note that it effervesces at once, giving off bubbles of carbon dioxide.   The mineral should not be powdered for the other rhombohedral carbonates will effervesce in the powdered form. When heated on charcoal, calcite does not fuse but carbon dioxide

Fig. 267.   Group of Calcite Crystals, Cumberland, England.

is driven off leaving calcium oxide.   In this way quicklime is made in the lime kiln to be used in making mortar.   The carbon dioxide driven off in the industrial process is utilized under pressure for many purposes; the most familiar is for charging carbonated drinks.

**Varieties.**   Limestone and Marble.   Calcite is an important rock-forming mineral for it is almost the only constituent in the limestones that underlie thousands of square miles of the earth's surface and in places are thousands of feet thick.   Shellfish, through the remote geologic past as well as those of the present day, secreted calcium carbonate from the sea water.   Their shells accumulate on the sea floor and on compaction may build up great thickness of limestone. When limestones are subjected to great pressures within the earth they may recrystallize, forming marble. Marble differs from limestone in texture, for in it one can see the cleavage faces on the coarser grains. Ordinary marble is white, but some varieties may be red, yellow, blue, or black and are then used for ornamental purposes.

Limestone finds its greatest use in the manufacture of cements and mortars and is thus the basic material of one of the great industries. It is used also as a building stone and as a flux in the smelting of iron ore. The lime used for agricultural purposes may be either quicklime or merely finely ground limestone.

Chalk, such as that found on the cliffs of Dover, England, is a fine-grained limestone made up of the tiny shells of marine organisms.

CAVE DEPOSITS. When water becomes charged with carbon dioxide, it has the power to dissolve limestones slowly as it works its way through them. In this way most of the limestone caves and caverns have formed. After formation, conditions may change so that the water, as it drips into the cave, may lose carbon dioxide and slowly deposit the dissolved calcium carbonate. Thus are built *stalactites*, which hang like icicles from the roof of the cavern, and *stalagmites*, which rise from the floor beneath. They are often very large and beautiful with a great variety of shapes. These deposits are often banded and in some places occur on a large scale, so that the rock can be quarried and used for ornamental stone. *Mexican onyx* is such a variety of calcite, delicate in coloring and beautifully translucent. Many caves with stalactites, stalagmites, and similar types of deposits are known in the United States. Probably the largest and most celebrated is Carlsbad Caverns, New Mexico.

SPRING DEPOSITS. Both hot and cold spring waters may deposit calcite on emerging from beneath the surface. Such deposits are known as *travertine* or *tufa*. The outstanding deposit of this type is at Mammoth Hot Springs, Yellowstone National Park, but small deposits of travertine are common in limestone regions.

SAND CRYSTALS. In certain places where water containing calcium carbonate in solution passes through sand, calcite may crystallize, incorporating the sand within it. Some crystals, such as those illustrated in Fig. 133, p. 52, may be made up of as much as 60 per cent quartz sand, but they still have the crystal form of calcite.

**Occurrence.** Calcite is such an abundant mineral that it is difficult in a limited space to list the outstanding occurrences. Limestone rocks are found in great quantities at many places in the world; in the United States they are especially common through the central states, Illinois, Iowa, Ohio, Wisconsin, etc. In cavities in the rocks the crystallized forms occur. Cavities, or geodes, lined with beautiful crystals are common. The tristate mining district of Missouri, Kansas, and Oklahoma should be mentioned in particular for the caves of this region abound in beautiful calcite crystals, some of which are very large weighing hundreds of pounds.

Calcite is found also as a gangue mineral associated with ores of metals such as lead, copper, and silver. The mines of the Lake Superior region have furnished many fine specimens, as have those of Derbyshire and elsewhere in England and the Harz region of Germany.

## DOLOMITE, $CaMg(CO_3)_2$

Dolomite is the rhombohedral carbonate intermediate between calcite and magnesite. It is a common and important mineral but is less abundant than calcite in both crystalline and massive forms. Rock dolomite is used both as a building and as an ornamental stone, as well as in the manufacture of certain cements. Recently it has been used as a raw material in the production of metallic magnesium. (See under magnesite, p. 194.)

**Habit.** Dolomite is rhombohedral, and crystals showing forms other than the unit rhombohedron are rare. The crystals have one peculiarity: the rhombohedron faces are almost always curved, giving a convex surface. The crystals of calcite are not curved but those of siderite, the iron carbonate, are. Some crystal aggregates are formed of many small crystals, and all are so curved that they have a saddle-shaped form, as illustrated in Fig. 269.

**Physical Properties.** Dolomite has perfect rhombohedral cleavage with angles so close to those of calcite that it is impossible to distinguish between them by inspection alone. It is a little harder than calcite, the hardness being $3\frac{1}{2}$–4, and it is a little denser, the specific gravity being 2.85. The luster is usually vitreous, but it is pearly in some varieties known as *pearl spar*. The color is commonly pale pink but may be colorless, white, gray, green, brown, or black.

**Composition.** Dolomite is the carbonate of calcium and magnesium, $CaMg(CO_3)_2$, with equal proportions of $CaCO_3$ and $MgCO_3$ in the pure mineral. Small amounts of iron commonly replace part of the magnesium; when considerable iron is present

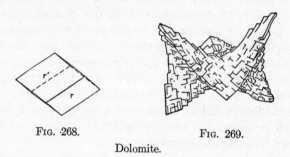

FIG. 268.　　　　FIG. 269.

Dolomite.

the mineral is called *ankerite*. Granular dolomite, unlike calcite, is not readily soluble in cold hydrochloric acid, but the powdered mineral will effervesce. This is the best way to distinguish between the two minerals.

**Occurrence.** The principal occurrence of dolomite is as a rock mineral, and as such it makes up tremendous masses of dolomitic limestones and crystalline dolomitic marbles. Much white marble is dolomite, not calcite. Most dolomite is believed to have been formed originally as limestone but has had part of the calcium replaced by magnesium to form dolomite. Great areas of the middle western states are underlain with dolomite. Crystals are common in the tristate mining district centering around Joplin, Missouri.

## MAGNESITE, $MgCO_3$

Magnesite is a mineral that has greatly increased in importance during the past half century. At the end of the nineteenth century it was described as a rare mineral; by 1930 it was no longer considered rare and was used in the production of magnesium oxide for refractory bricks and insulating purposes. It was used also in the preparation of magnesium salts for medicines and paper manu-

facture.    Because carbon dioxide is given off at a lower temperature
than it is from limestone, magnesite is used as a source of this gas.
Today, besides its former uses, it is an ore of magnesium.

Magnesium as a metal has been known for many years, but it
was produced on an extremely small scale from magnesium chloride,
recovered from the brines of salt wells at Midland, Michigan.
Magnesium is a lighter metal than aluminum but resembles it in
many of its other properties.    With the increased demand for
lighter alloys for the tremendous production of aircraft for World
War II, new raw materials were sought as a source of magnesium.
Thus magnesite and, to a lesser degree, dolomite have become ores
of magnesium.    One of the most interesting developments in
magnesium production has been the extraction of this metal from
sea water.    In Texas a gigantic plant has been constructed that
handles millions of gallons of sea water a day and from it separates
the small percentage of magnesium chloride present to be later
converted into metallic magnesium.

Because of the present great capacity for magnesium production,
we may expect in the years to come that more and more articles
will be made of magnesium and its alloys, in place of the heavier
metals.

**Habit.**    Magnesite is rhombohedral, but, unlike calcite and
dolomite, crystals are rarely seen.    It is usually in compact masses
and less frequently in granular masses showing definite cleavage.

**Physical Properties.**    Magnesite has perfect rhombohedral cleavage with a cleavage angle of 72° 36'.    The hardness is $3\frac{1}{2}$–5, the
compact massive variety having the greater hardness.    The
specific gravity is 3–3.2.    The luster is vitreous in crystals and dull
in the massive variety.    The color of pure magnesite is white,
but impurities may color it gray, yellow, or brown.

**Composition.**    Magnesium carbonate, $MgCO_3$.    It is almost
insoluble in cold acid but dissolves with effervescence in hot hydrochloric acid.

**Occurrence.**    Magnesite occurs in two distinct types, one fine-
grained and compact, the other coarse and cleavable.    The compact variety is found in irregular veins and masses in serpentine
rocks from which it has been derived through the action of waters
containing carbon dioxide.    This type has been mined on the
Island of Euboea, Greece, and in the serpentine of the Coast
Range of California.    The cleavable variety occurs in sedimentary

strata and appears to have formed by the complete substitution of magnesium for the original calcium of the limestone. The most famous deposit of this type is at Styria, Austria. In the United States large masses are found in Stevens County, Washington.

## SIDERITE, $FeCO_3$

Siderite, or *spathic ore*, as it is frequently called, is in certain places used as an ore of iron, but it is much less important than the three oxides. It is mined in Great Britain and Austria but is very subordinate in the United States.

**Habit.** Siderite is rhombohedral, and crystals usually show the unit rhombohedron with curved surfaces resembling the crystals of dolomite. It is found usually in cleavable aggregates but may also be concretionary or in botryoidal or earthy masses.

**Physical Properties.** Siderite has perfect rhombohedral cleavage with a cleavage angle of 73°. The hardness is $3\frac{1}{2}$–4; the specific gravity, 3.85. The luster is vitreous, and the color is light to dark brown.

**Composition.** Siderite is ferrous carbonate, $FeCO_3$. The 48.2 per cent iron present is a much lower percentage than that in the iron oxides. When siderite is heated on charcoal, it becomes strongly magnetic and turns black. It dissolves with effervescence in hot hydrochloric acid.

**Occurrence.** Siderite may be formed by the action on limestone of solutions carrying iron. If replacement of calcium by iron is extensive, large bodies of ore may be formed, as at Styria, Austria. Concretionary siderite, known as *clay ironstone*, is found in extensive beds in England, where it has been mined. Similar deposits are known in western Pennsylvania and eastern Ohio, but they have rarely been mined as iron ore. Siderite is found also in crystals in veins associated with metallic ores.

## RHODOCHROSITE, $MnCO_3$

**Habit.** Rhodochrosite is rhombohedral, and crystals, although not common, are found in places in beautiful clear rhombohedrons. Some crystals may show curved surfaces. It is usually in cleavable masses.

**Physical Properties.** Rhodochrosite has perfect rhombohedral cleavage with a cleavage angle of 73°. The hardness is $3\frac{1}{2}$–$4\frac{1}{2}$; the specific gravity, 3.45–3.6. The luster is vitreous, and the color is some shade of rose-red. Some varieties may be very pale pink, and others dark brown.

**Composition.** Manganese carbonate, $MnCO_3$. A small amount of iron may be present. Rhodochrosite dissolves in hot hydrochloric acid with effervescence. When it is fused with sodium carbonate, the mass assumes a blue-green color indicating manganese.

**Occurrence.** Rhodochrosite is found frequently as a gangue mineral in veins carrying silver, lead, and copper ores. In certain veins at Butte, Montana, it is so abundant that it is mined as an ore of manganese. Beautiful clear crystals of a deep rose-red color come from Lake County, Colorado.

### SMITHSONITE, $ZnCO_3$

Smithsonite is sometimes called *dry-bone ore* by the miners because of the cellular texture of certain specimens that resemble bone. It is used as a zinc ore but is considerably subordinate to the sulfide, sphalerite. In some places smithsonite is found in translucent green or greenish blue masses which are cut and polished for ornamental purposes.

**Habit.** Like the other members of the calcite group, smithsonite is rhombohedral but is rarely found in crystals. It usually forms botryoidal and stalactic masses, or honeycombed masses, known as dry-bone ore.

**Physical Properties.** Smithsonite has perfect rhombohedral cleavage, but, because of the nature of the masses in which it occurs, the cleavage is seldom seen. The hardness is 5; the specific gravity, 4.4, unusually high for a carbonate. The luster is vitreous, and the color is usually dirty brown but may be white, green, blue, or pink. A yellow variety is known as *turkey-fat ore* and contains cadmium.

**Composition.** Zinc carbonate, $ZnCO_3$. Smithsonite resembles calcite in that it effervesces in cold hydrochloric acid but can be distinguished from calcite by its high specific gravity.

**Occurrence.** Smithsonite is a secondary mineral formed by the alteration of sphalerite and is found in the upper portions of zinc deposits. Certain types, particularly *dry-bone ore*, are difficult to recognize, and it has not been uncommon in mining operations to overlook smithsonite as an ore mineral. Fine specimens from Laurium. Greece, and Kelley, New Mexico, have been cut for

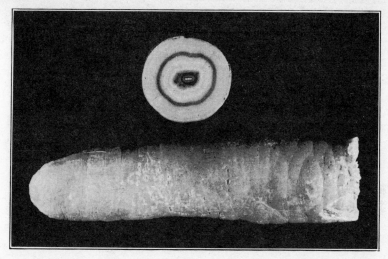

FIG. 270. Smithsonite Stalactite and Cross Section, Sardinia.

ornamental purposes. In the United States it has been mined as an ore of zinc in Arkansas, Missouri, Wisconsin, and Colorado.

## ARAGONITE, $CaCO_3$

Aragonite is a dimorphic form of calcium carbonate but is much less important than calcite. It heads the list of the orthorhombic

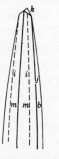

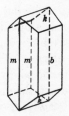

FIG. 271.          FIG. 272.

Aragonite Crystals.

carbonates known as the *aragonite group*. This group includes aragonite, strontianite, witherite, and cerussite, all of which are closely related in their crystal and physical properties.

**Habit.**   Aragonite is orthorhombic with slender pointed crystals commonly in radiated groups (Fig. 275).   It is found also in twin crystals resembling a hexagonal prism with basal pinacoid (Fig. 273).   Such pseudohexagonal crystals are characteristic of all the members of the aragonite group.   Besides these other types occur,

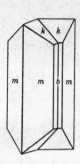

Fig. 273.                              Fig. 274.

Aragonite Twins.

such as the delicate coral-like *flos ferri* (Fig. 161, p. 62) found in some iron mines.   It is found also in reniform and massive aggregates.

**Physical Properties.**   There are two poor cleavages, one parallel to the side pinacoid and the other parallel to the prism.   In single crystals or columnar aggregates, therefore, all the cleavages are parallel to the length.   This may be confused with columnar calcite which also has cleavages parallel to the length of the individuals, but calcite has another cleavage plane at the end of the columns, whereas aragonite terminates in an irregular fracture.

The hardness is 3½–4; the specific gravity, 2.95—both greater than those of calcite.   The luster is vitreous.   Aragonite may be colorless, white, or pale yellow.

**Composition.**   Calcium carbonate, $CaCO_3$, the same as calcite. It effervesces readily in cold hydrochloric acid and gives the same reaction as calcite to all other chemical tests.

**Occurrence.**   Aragonite is much less common than calcite and less stable once it is formed.   Thus pseudomorphs of calcite after aragonite are common.   A good example is the calcium carbonate secreted by mollusks.   The mother-of-pearl on the inside of the shell is aragonite, but it is changed to calcite on the outside.

Well-formed pseudohexagonal crystals are found in Aragon.

Spain, whence the mineral derives its name. The best-developed pointed type of crystals is found in England at Alston Moor and Cumberland. From the iron mines in Austria come the best specimens of flos ferri. In the United States this variety is found in the Organ Mountains, New Mexico, and at Bisbee, Arizona.

FIG. 275. Aragonite Crystals, Cumberland, England.

## WITHERITE, $BaCO_3$

Witherite, barium carbonate, is a rare mineral compared to barite, barium sulfate. However, in certain places it is abundant enough to become a minor source of barium, as at El Portal, Yosemite National Park, California.

**Habit.** Witherite is orthorhombic, and crystals resemble those of aragonite; that is, they are found in both acicular radiating groups and as pseudohexagonal twins. These twins form six-sided pyramids having somewhat the aspect of quartz (Fig. 276). Massive witherite also occurs.

**Physical Properties.** There is good prismatic cleavage; thus the cleavage is parallel to the long dimension of the crystals. The hardness is $3\frac{1}{2}$; the specific gravity, 4.3, unusually high for a nonmetallic mineral, and easily detected by picking up a specimen. The luster is vitreous, and the color is white to gray.

**Composition.** Barium carbonate, $BaCO_3$. When a fragment is fused, it imparts a green color to the flame; this is the test for barium. Like calcite and aragonite, it is soluble in cold hydrochloric acid with effervescence. It can be distinguished from these minerals by its high specific gravity.

## STRONTIANITE, $SrCO_3$

Strontianite, together with the strontium sulfate, celestite, is the source of the element strontium. Strontium has no great commercial significance but is used in fireworks and in the separation of sugar from molasses.

**Habit.** Strontianite is orthorhombic. Crystals are usually radiating acicular like those of aragonite, but strontianite is also found in pseudo-hexagonal twins. It may be columnar, fibrous, or granular.

**Physical Properties.** Strontianite like witherite has good prismatic cleavage. The hardness is $3\frac{1}{2}$–4; the specific gravity, 3.7. The specific gravity is lower than witherite but still high for a nonmetallic mineral. The luster is vitreous, and the color is white, pinkish, gray, yellow, or green.

FIG. 276. Witherite Twin

**Composition.** Strontium carbonate, $SrCO_3$. A fragment will not fuse but will usually fly to pieces when heated, giving a red color to the flame. This is the best test to distinguish it from witherite, for, like witherite, it is soluble in cold hydrochloric acid.

**Occurrence.** Strontianite is a comparatively rare mineral, less abundant than celestite. It is commonly found in veins in limestone. It has been mined in Westphalia, Germany.

## CERUSSITE, $PbCO_3$

**Habit.** Cerussite is orthorhombic. Although it belongs to the aragonite group, its crystals are somewhat different from those of the other members. They are often in tabular plates (Fig. 277) or twinned with plates crossing each other at 60° angles (Fig. 279). Another expression of the same type of twinning is pseudo-hexagonal pyramids as shown in Fig. 278. Cerussite is also found in granular massive aggregates.

**Physical Properties.** Cerussite has prismatic cleavage. The hardness is $3$–$3\frac{1}{2}$; the specific gravity is 6.55, high foɪ a nonmetallic mineral but to be expected in a lead compound. The luster is adamantine. Crystals are frequently colorless, but massive aggregates are usually white or gray. The clear and colorless crystals do not perhaps at first suggest to the eye that the mineral contains lead. A more careful examination, however, shows the adamantine luster possessed by crystals that strongly refract light. A high index of refraction, $\varepsilon$ ‹ this property is called, is usually found in very hard minerals (as in diamond) or in those

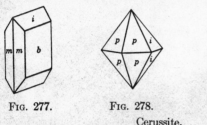 

Fig. 277.  Fig. 278.  Fig. 279.

Cerussite.

containing heavy atoms, like lead. All the compounds of lead have an adamantine or resinous luster, as does the lead glass, called paste, of which imitation gems are made.

**Composition.** Cerussite is lead carbonate, $PbCO_3$. When mixed with sodium carbonate and heated on charcoal, it is reduced to a lead globule. Cerussite is the only member of the aragonite group that is not soluble in hydrochloric acid, but it will dissolve in warm dilute nitric acid with effervescence.

**Occurrence.** Cerussite, next to galena, is the commonest ore of lead. The two minerals are frequently associated, for cerussite is produced in nature's laboratory by the action of carbonated water on galena. Not uncommonly cerussite is found with a core of galena. Cerussite, therefore, is a secondary mineral found in the upper portion of lead deposits, and mines that are worked for cerussite in the upper levels are usually worked for galena in depth. One of the best examples is at Broken Hill, New South Wales. In the United States beautiful crystals come from the Organ Mountains, New Mexico. It is found also in Arizona, Colorado, and Idaho.

## MALACHITE, $Cu_2CO_3(OH)_2$

Malachite, known also as *green copper carbonate*, is usually associated with the blue copper carbonate, azurite. Specimens

FIG. 280.   Cerussite, Reticulated Aggregate, Tsumeb, Southwest Africa.

containing the two minerals are frequently outstandingly beautiful because of their striking colors.   Malachite in addition to being an ore of copper is used to some extent for ornamental purposes. The Russians have been particularly skillful in making art objects of it.   Thin slices are cut and used as a veneer on table tops, vases, and pillars in churches.   Most of the choice Russian malachite pieces have been brought together in the Hermitage in Leningrad and are on exhibit there.

**Habit.**   Malachite is monoclinic, but crystals are rare and indistinct.   It is usually in radiating fibers that form botryoidal masses (Fig. 281).   Well-formed crystals of malachite are commonly pseudomorphs after azurite.

**Physical Properties.**   The perfect basal cleavage of malachite is rarely seen because of the fibrous nature of the mineral.   The hardness is $3\frac{1}{2}$–4, and the specific gravity is about 4.   The luster is silky in fibrous varieties but vitreous in crystals.   The color is bright green.

**Composition.**   A basic carbonate of copper, $Cu_2CO_3(OH)_2$ Before the blowpipe malachite fuses, giving a green flame.   It dissolves with effervescence in cold hydrochloric acid and, in this way, can be distinguished from other green copper minerals.

**Occurrence.** Both malachite and azurite are copper ores of secondary origin; that is, they have formed near the surface by the action of carbonated water on the primary copper minerals, such as chalcopyrite and bornite. They are thus associated with

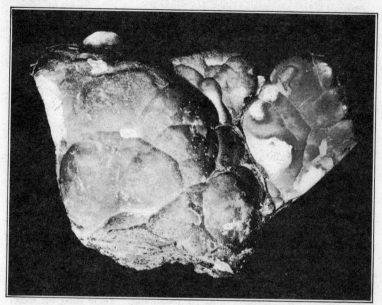

FIG. 281. Botryoidal Malachite, Bisbee, Arizona.

minerals of similar origin, such as cuprite, native copper, and iron oxide. Fine specimens have come from Tsumeb, Southwest Africa; Chessy, France; and South Australia. In the United States the outstanding locality is Bisbee, Arizona.

### AZURITE, $Cu_3(CO_3)_2(OH)_2$

**Habit.** Azurite is monoclinic and, unlike malachite, is frequently found in well-formed crystals. It may also be in radiating groups.

**Physical Properties.** The intense azure-blue color of azurite is its most striking property but well-formed crystals may appear almost black. The hardness is $3\frac{1}{2}$–4, and the specific gravity is 3.77. The luster is vitreous.

**Composition.** Azurite is a basic copper carbonate, $Cu_3(CO_3)_2(OH)_2$, very similar to malachite, but containing slightly less water. It effervesces in cold hydrochloric acid, and other tests are the same as those for malachite.

**Occurrence.** See under malachite.

## BORATES

There are over a score of borates, but only four of them are abundant enough to warrant description in this book. They are borax, kernite, ulexite, and colemanite. Each of them has at one time been the chief source of the final commercial product, borax.

### Borax, $Na_2B_4O_7 \cdot 10H_2O$

The medicinal qualities of borax have been known for many centuries, and for this reason it was early imported into Europe. The original source was Tibet. Later it was obtained by the

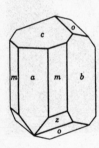

FIG. 282. Borax.

evaporation of water from hot springs in Tuscany in northern Italy. In 1860 it was discovered in Lake County, California, and since that time, although the source has shifted from one place to another, the United States has been the chief producer.

Borax is still used in medicine and as an antiseptic, but its chief use today is in washing powder and in cleansing. It is also an excellent solvent for metallic oxides and thus finds a use in certain smelting operations. The mineralogist uses it in a similar manner when he dissolves a mineral fragment in a borax bead to obtain the color imparted by certain elements.

**Habit.** Borax is monoclinic, and prismatic crystals (Fig. 282) have been common at certain localities. It is found also in porous masses.

**Physical Properties.** Borax has perfect cleavage parallel to the front pinacoid. The hardness is $2-2\frac{1}{2}$; the specific gravity is 1.7 (very low). It is colorless in fresh crystals, but the crystals lose water and turn white on standing in a dry atmosphere. It is soluble in water and has a sweetish-alkaline taste.

**Composition.** Hydrous sodium borate, $Na_2B_4O_7 \cdot 10H_2O$. It is easily fusible and imparts a strong yellow color to the flame (sodium). If moistened with sulfuric acid, it gives a green color to the flame indicating boron. In the laboratory colorless borax crystals lose five molecules of water and become the chalky white substance, *tincalconite*, $Na_2B_4O_7 \cdot 5H_2O$.

**Occurrence.** Borax was a relatively rare mineral until the middle of the nineteenth century when it was discovered in California. It was first found in Lake County and later in Death Valley, Inyo County. The famous twenty-mule team was used to haul borax from Death Valley, 135 miles to the railroad to the south.

## KERNITE, $Na_2B_4O_7 \cdot 4H_2O$

Kernite, a relatively new mineral, was first described in 1926. At the present time its only occurrence is on the Mohave Desert at Kramer, California. The new minerals that are described from time to time are usually found sparingly, some yielding barely enough material for an adequate description. Kernite was an exception to the rule, for it was found by thousands of tons and, moreover, became almost immediately a valuable commercial mineral. Today about half of the borax produced in the United States has its source in kernite. If one compares the formulas of borax and kernite, one finds that the only difference is the higher percentage of water in borax. Kernite is, therefore, a remarkable mineral, for a pound of it will yield about 1.4 pounds of borax, the finished product!

**Habit.** Kernite is monoclinic. It is usually found in coarse cleavable aggregates.

**Physical Properties.** Perfect cleavage parallel to both the basal pinacoid and the front pinacoid gives rise to splintery fragments elongated parallel to the *b* crystallographic axis. The hardness is 3; the specific gravity, 1.95. The luster is vitreous to pearly. Kernite is usually colorless when fresh, but such specimens become coated with the chalky white mineral tincalconite on long exposure to the air.

**Composition.** Hydrous sodium borate, $Na_2B_4O_7 \cdot 4H_2O$. Kernite fuses easily to a clear glass. Like borax, it is soluble in cold water.

## ULEXITE, $NaCaB_5O_9 \cdot 8H_2O$

Ulexite has from time to time served as a source of borax. It characteristically forms *cotton-balls* (a name frequently applied to it) on the surface of playa lakes where it has crystallized from brines.

**Habit.** Ulexite is triclinic. Crystals are found only as fine silky fibers, aggregates of which usually form rounded masses to which the name cotton-balls is given.

**Physical Properties.** The hardness is 1; the specific gravity, 1.96; the luster is silky; the color, white.

**Composition.** Ulexite is a hydrous sodium and calcium borate, $NaCaB_5O_9 \cdot 8H_2O$. It fuses easily to clear glass, coloring the flame deep yellow. If it is moistened with sulfuric acid and then introduced into the flame, a momentary flash of green is seen, indicating boron.

**Occurrence.** Ulexite is usually associated with borax and is found on the surface of dry lakes in enclosed basins. In such a manner it has been found in California and Nevada, as well as in Chile and Argentina.

## COLEMANITE, $Ca_2B_6O_{11} \cdot 5H_2O$

From 1886 when colemanite was found at Monte Blanco, on the rim of Death Valley, until the discovery of kernite in 1926, colemanite was a major source of borax, furnishing over half of the world's supply.

**Habit.** Colemanite is monoclinic. Crystals are usually short prismatic. The mineral is most commonly found in cleavable masses.

**Physical Properties.** A perfect cleavage parallel to the side pinacoid is colemanite's outstanding property. The luster is vitreous, particularly well seen on the cleavage face. The hardness is 4–4½; the specific gravity, 2.42. Colemanite is colorless to white.

**Composition.** Hydrous calcium borate, $Ca_2B_6O_{11} \cdot 5H_2O$. When a fragment is held in the flame, it fuses easily and crumbles, imparting a green color to the flame (boron).

**Occurrence.** Colemanite has been found only in California and Nevada but it occurs in several places there and has been mined as a source of borax.

## PHOSPHATES

The phosphates form a large group of minerals, but most of them are very rare and are not considered here. Mimetite, an arsenate, and vanadinite, a vanadate, are included with the phosphates.

## APATITE, $Ca_5(F,Cl)(PO_4)_3$.

Among the large group of phosphates, apatite is the only one that can be considered a common mineral; many of the others have resulted from its alteration. In certain rare instances apatite is found in beautifully colored transparent crystals and is cut as a gemstone. However, commercially much greater importance attaches to apatite as a source of phosphorus, a constituent of most commercial fertilizers, and in some places apatite is mined as a source. Today, however, sedimentary beds known as phosphate rock or phosphorite are most extensively mined as a source of this element. After being mined, both apatite and rock phosphate are treated with sulfuric acid to make superphosphate, for in this form they are much more soluble in the dilute acids of the soil.

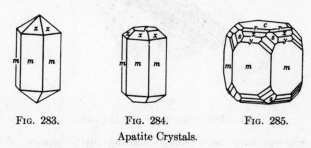

FIG. 283.  FIG. 284.  FIG. 285.

Apatite Crystals.

**Habit.** Apatite is hexagonal, and crystals are common showing prism, base, and pyramids. (See Figs. 283, 284.) It is also found in granular and compact masses.

**Physical Properties.** On some crystals a poor cleavage may be seen parallel to the basal pinacoid. The hardness is 5; apatite can just be scratched by the knife. The specific gravity is 3.15–3.20. The color is usually green or brown, but it may be violet, blue, yellow, or colorless. The luster is vitreous.

**Composition.** Apatite is usually calcium fluophosphate, $Ca_5F(PO_4)_3$ but may contain chlorine instead of fluorine.

**Occurrence.** Apatite is widespread and is found as an accessory mineral in tiny isolated grains in all types of rocks. Gem apatite of a beautiful purple color has been found at Auburn, Maine. At one time apatite was mined in Ontario, Canada, where it occurred in crystals and masses in coarsely crystalline calcite (Fig. 286). Undoubtedly the largest known mass of apatite is in the mountains of the Kola peninsula in northwest Russia. Here, in a granular form, it occurs intimately mixed with nepheline. Since 1930 it has been mined extensively as a source of phosphorus.

It is largely from the chemical breakdown of apatite that the phosphorus normally used in plant life is contributed to the soil. Some of this phosphorus, however, is carried away in solution and eventually reaches the ocean where it may be extracted by marine organisms. In certain places on the sea floor organic remains rich

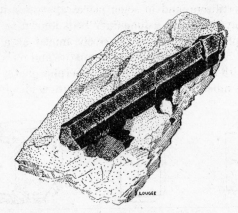

Fig. 286.   Apatite Crystal in Calcite.

in phosphorus may accumulate building up a deposit of *rock phosphate.*

Rock phosphate or *phosphorite* has as its principal constituent an amorphous material, *collophanite.* The formula for collophanite has been written as $Ca_3(PO_4)_2 \cdot H_2O$, but it usually contains varying amounts of other elements.

Bone is calcium phosphate, and deposits of phosphorite are formed by the accumulation of animal remains. In the United States large deposits of phosphate rock are found in Wyoming, Idaho, Tennessee, and Florida. Some of the world's largest deposits are located in North Africa, in Tunisia, Algeria, and Morocco.

## PYROMORPHITE, $Pb_5Cl(PO_4)_3$

Pyromorphite, mimetite, and vanadinite are closely related both in chemical composition and crystal form. All three are secondary minerals formed in the upper oxidized portion of lead veins. Pyromorphite is the most abundant and in places has been mined as an ore of lead.

**Habit.** Pyromorphite is hexagonal and is found in small hexagonal prisms which are frequently cavernous and may show rounded or barrel-shaped forms. The crystals may be clustered together branching out from a slender stem, as shown in Fig. 287. It occurs also as a thin crust or coating, which may be covered with tiny crystals or simply globular.

**Physical Properties.** The hardness is $3\frac{1}{2}$–4, and, like all lead compounds, it has a high specific gravity, viz., 6.5–7.1. The luster is resinous, and the color is commonly green, varying from grass-green to both lighter and darker shades; it may also be pale brown. The streak is white even in the deep green varieties.

Fig. 287. Pyromorphite.

Fig. 288. Pyromorphite Crystals.

**Composition.** Pyromorphite is lead chlorophosphate, $Pb_5Cl(PO_4)_3$. Some arsenic may be present. It fuses easily on charcoal. If the fragment is examined after it is completely fused, it will be seen that it is nearly spherical and sparkles on the surfaces from the reflection of light from a multitude of facets. Hence the name of the mineral comes from the Greek words meaning *fire* and *form*. If the fused globule is further heated on charcoal with the addition of sodium carbonate, it yields a globule of metallic lead.

**Occurrence.** In the United States pyromorphite has been found at Phoenixville, Pennsylvania; Davidson County, North Carolina; and in Idaho.

## MIMETITE, $Pb_5Cl(AsO_4)_3$;  VANADINITE, $Pb_5Cl(VO_4)_3$

As already mentioned, these two minerals are closely related to pyromorphite in composition and form. In fact mimetite is named from the Greek meaning *imitator* because of its close resemblance to pyromorphite, although crystals are as a rule rarer and less distinct.

MIMETITE has a hardness of $3\frac{1}{2}$ and a specific gravity of 7–7.2. It may be colorless, yellow, orange, or brown with a resinous or adamantine luster. It is essentially lead chloroarsenate, $Pb_5Cl(AsO_4)_3$, but some phosphorus may be present. A lead globule results, if mimetite is heated on charcoal with sodium carbonate.

FIG. 289.  Vanadinite

It is easily recognized by the arsenical fumes with tne garlic odor that are yielded when it is heated.

VANADINITE is often a fine deep red color, and when crystals are sharp and clear it is one of the most beautiful minerals. Less brilliant yellow and light brown varieties also occur. The crystals are hexagonal prisms terminated by the base and more rarely by hexagonal pyramids. Cavernous forms also occur as with pyromorphite. The hardness is 3, and the specific gravity is 6.7–7.1. The composition is $Pb_5Cl(VO_4)_3$; usually small amounts of arsenic and phosphorus are present. The reactions for lead on charcoal are like those of pyromorphite.

Besides being a minor ore of lead, vanadinite is also a source of the rare element vanadium. This element is used chiefly for hardening steel, but is used also for fixing the colors in dyeing fabrics and as a yellow pigment in the form of metavanadic acid.

## AMBLYGONITE, $LiAlFPO_4$

Amblygonite is a rare mineral that in a few places is found abundantly enough to be mined as a source of lithium.

**Habit.** Amblygonite is triclinic, but crystals are small and rare. It is usually found in coarse cleavable masses.

**Physical Properties.** Amblygonite has perfect cleavage parallel to the base and a poorer cleavage parallel to the front pinacoid. These two cleavages are nearly at right angles and thus resemble the cleavages of feldspar. The hardness is 6, and the specific gravity 3.0–3.1. The color is white, pale green, or blue. The luster is pearly on the base; elsewhere it is vitreous.

**Composition.** Lithium aluminum fluophosphate, $LiAlFPO_4$. Amblygonite fuses easily before the blowpipe, yielding a red flame. By this test one can distinguish it from feldspar, with which it may be confused.

**Occurrence.** Amblygonite is found in pegmatites where it is associated with other lithium-bearing minerals, such as lepidolite and spodumene. It is found at various localities in Maine; at Pala, California; and in the Black Hills, South Dakota.

## WAVELLITE, $Al_3(OH)_3(PO_4)_2 \cdot 5H_2O$

**Habit.** Wavellite is orthorhombic; crystals are rare; and it is usually in radiating globular aggregates. (See Fig. 156, p. 60.)

**Physical Properties.** Wavellite has good cleavage parallel to the side pinacoid and macrodome. The hardness is $3\frac{1}{2}$–4; the specific gravity is 2.33. The luster is vitreous, and the color is green, yellow, white, or brown.

**Composition.** Wavellite is a hydrous basic aluminum phosphate, $Al_3(OH)_3(PO_4)_2 \cdot 5H_2O$. It is infusible but when heated it swells and falls apart into fine particles.

**Occurrence.** Wavellite is a rare mineral formed by secondary processes, the phosphorus probably being derived from the alteration of apatite. It is found in several localities in Pennsylvania and Arkansas.

## TURQUOIS,   $Al_2(OH)_3PO_4 \cdot H_2O + Cu$

Turquois is a semi-precious stone that has been used for ornamental purposes for centuries.  The early material came from Persia and was carried through Turkey to reach Europe.  The mineral derives its name from the French for *Turkish.*

**Habit.**   Turquois is triclinic but is rarely in minute crystals.  It is usually cryptocrystalline, forming compact reniform masses. It occurs also in thin veins and as crusts.

**Physical Properties.**   The bluish green color is the outstanding property and the one that makes it desirable as a gem material. The hardness is 6; the specific gravity, 2.6–2.8.  The luster is waxlike.

**Composition.**   A basic hydrous phosphate of aluminum, $Al_2(OH)_3PO_4 \cdot H_2O + Cu$.  When touched with a drop of hydrochloric acid and heated, it gives a blue flame (test for copper).

**Occurrence.**   Like many of the other phosphates, turquois is of secondary origin.  It is found usually in small veins cutting volcanic rocks.  The famous Persian deposits in the province of Khorasan are still producing.   In the United States it is found in Arizona, Nevada, and California; but the most famous locality is near Santa Fe, New Mexico.   There the Navajo Indians obtain the turquois which they mount in jewelry.

## SULFATES

There are many sulfate minerals, but only the few that are common are considered here.   The anhydrous sulfates described are barite, celestite, anglesite, and anhydrite, which form a closely related group with similar crystallographic and physical properties. The only hydrous sulfates described are gypsum and alunite.

### BARITE,   $BaSO_4$

The outstanding property of barite is its high density; because of its density it is often called *heavy spar.*   As it is the most common and widespread barium mineral, it is the chief source of that element.   The carbonate, witherite, is the only other important barium mineral.   The principal use of barium is in the substance *lithopone,* a mixture of barium sulfide and zinc sulfate, which is used in paints and in the manufacture of linoleum and textiles.   Much of the barite mined is ground to a fine powder and made into a sludge of high specific gravity for use in drilling deep wells.

**Habit.** Barite is orthorhombic. Crystals are common; the usual habit is tabular parallel to the base. (Fig. 290.) Some of the other modifying forms are shown in Figs. 291 to 295. Ag-

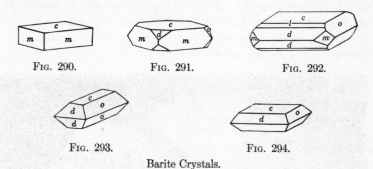

FIG. 290.          FIG. 291.                    FIG. 292.

FIG. 293.                    FIG. 294.

Barite Crystals.

gregates of divergent plates frequently form *crested barite* or *barite roses*, as shown in Fig. 296. Barite occurs also in massive or

*Ward's Natural Science Establishment*

FIG. 295.   Barite Crystals, Cumberland, England.

granular aggregates, and in such forms it is most difficult to recognize.

**Physical Properties.** Barite has both basal and prismatic cleavage. Cleavage fragments at first glance may resemble those of calcite for the angles between the prism faces are close to the rhombohedron angle of calcite. On comparing these two minerals, the student will find that the basal cleavage is at right angles to the prismatic cleavage in barite, whereas in calcite the third cleavage is not at right angles to the other two.

Fig. 296. Crested Barite.

The hardness is $3-3\frac{1}{2}$ and the specific gravity is 4.5, extremely high for a nonmetallic mineral. Barite may be colorless, white, or light shades of blue, yellow, and red. The luster is vitreous.

**Composition.** Barite is barium sulfate, $BaSO_4$. When heated before the blowpipe, it yields a yellowish green flame (test for barium). It can be distinguished from witherite, which gives the same flame test, by its insolubility in hydrochloric acid.

**Occurrence.** Barite is often found as a gangue mineral associated with lead, copper, and silver ores. It is found also in veins in limestone or in residual masses in clay overlying limestones. Beautiful crystals have come from several localities in England chiefly Derbyshire, Westmoreland, Cornwall, and Cumberland. In the United States they have been found at Cheshire, Connecticut; Dekalb, New York; and Fort Wallace, Idaho. Massive barite has been mined in Georgia, Tennessee, Missouri, Arkansas, and California.

## CELESTITE, SrSO₄

**Habit.** Celestite is orthorhombic, and the crystals resemble so closely those of barite that it is difficult to distinguish between them without careful angular measurements. Crystals are commonly tabular but may be elongated parallel to the *a* axis. (See Fig. 298.) It occurs also in granular and fibrous aggregates.

FIG. 297.   FIG. 298.

Celestite Crystals.

**Physical Properties.** Perfect cleavage parallel to the base and prism yields fragments closely resembling those of barite. The hardness is 3–3½; the specific gravity, 3.95–3.97 (although high for a nonmetallic mineral, it is less than that of barite). The luster is vitreous, but it may be pearly on the base. The color is commonly white, but the crystals, as well as the fibrous forms, often show a tinge of blue, and, although this is not an essential characteristic, the blue tinge is so common that it has given to the species the name derived from the Latin *coelestis*.

**Composition.** Strontium sulfate, SrSO₄. A fragment heated in the forceps colors the flame a deep red (test for strontium). Celestite's insolubility in hydrochloric acid distinguishes it from the soluble strontium carbonate, strontianite.

**Occurrence.** Celestite, both as disseminations and as the linings of cavities, is usually associated with limestone. It is found also in sandstone, gypsum, rock-salt, and clay. In the United States it is found in West Virginia, Texas, and California. A most striking occurrence is on the Island of Put-in-Bay in Lake Erie. Here one can today descend into a cave that is completely lined with celestite crystals, some measuring more than a foot in length.

## Anglesite, $PbSO_4$

**Habit.** Anglesite is orthorhombic, and crystals resemble in form and angles those of barite and celestite but are usually of a more varied and complex development. It is found also in granular or compact masses and may show concentric banding around a core of galena.

**Physical Properties.** Anglesite has imperfect cleavage parallel to the prism and base. The hardness is 3; the specific gravity, 6.2–6.4. The luster is adamantine when the specimen is crystalline, but it may be dull when the specimen is impure and earthy. The crystals are usually clear and colorless, but massive material may be gray, pale shades of yellow, or brown.

**Composition.** Anglesite is lead sulfate, $PbSO_4$. On charcoal before the blowpipe it decrepitates and fuses readily to a clear bead, which becomes milk-white on cooling. When heated on charcoal with soda, it yields a bead of metallic lead. It dissolves with difficulty in nitric acid and does not effervesce as does cerussite; hence the two minerals are easily distinguished.

**Occurrence.** Anglesite, like cerussite, is a secondary mineral formed by the oxidation of galena. It is consequently found near the surface of lead veins and associated with other oxidized minerals, such as cerussite, smithsonite, and iron oxides. It derives its name from the original locality on the Island of Anglesey, Wales. Other world-famous localities are Broken Hill, New South Wales; Monte Poni, Sardinia; and Otari, Southwest Africa. In the United States it is found at Phoenixville, Pennsylvania; Tintic District, Utah; and the Coeur d'Alene District, Idaho.

## Anhydrite, $CaSO_4$

Anhydrite receives its name because, although consisting of calcium sulfate, it does not, like gypsum, contain water. It is less common than gypsum and has little industrial use.

**Habit.** Anhydrite is orthorhombic, but crystals are rare and it is usually in cleavable masses. It may also be massive, fibrous, or granular.

**Physical Properties.** Anhydrite has distinct cleavage parallel to the three pinacoids. The fragments may, therefore, appear cubic, but careful examination shows that the three cleavages are unlike as they must be since anhydrite is in the orthorhombic system. The hardness is 3–3½ and the specific gravity is 2.89–2.98. The color is usually white or gray but may have a blue or red tinge.

**Composition.** Calcium sulfate, $CaSO_4$. Unlike gypsum it contains no water.

**Occurrence.** Anhydrite is much less common than gypsum but occurs in much the same manner and with the same mineral association, with bedded salt deposits and limestone.

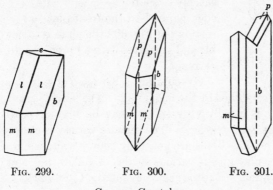

| Fig. 299. | Fig. 300. | Fig. 301. |

Gypsum Crystals.

## GYPSUM, $CaSO_4 \cdot 2H_2O$

Gypsum is a common mineral and of considerable commercial importance because of its use in the production of plaster of Paris. This material is made by heating ground gypsum until about three-fourths of the water is driven off. When the plaster of Paris is mixed with water, it slowly crystallizes and hardens, assuming the shape of the confining surfaces. It is thus used to make casts and molds of all kinds. Its most extensive use is in the building industry in the form of gypsum lath, wall board, and plaster for interior use.

*Alabaster* is a fine-grained massive variety of gypsum that is cut and polished for ornamental purposes.

**Habit.** Gypsum is monoclinic, and crystals often have the habit of those illustrated in Figs. 299, 300, and 303. Twin crystals are also common, especially those of the "swallow-tail" type, like that in Fig. 301. It is found also in cleavable and fine granular masses. A fibrous variety with a silky luster is known as *satin spar*. The fine-grained massive variety as mentioned above is *alabaster*.

**Physical Properties.** The crystals have very perfect cleavage parallel to the side pinacoid, and sometimes very large, thin, and perfectly transparent plates may be obtained. The variety yielding these is called *selenite*. The plates look a little like mica but are much softer and, though somewhat flexible, are quite inelastic. When broken carefully, a plate shows two other directions of cleavage. In one of these directions a plate breaks rather sharply, "snap cleavage," with a conchoidal edge parallel to the front pinacoid (*a*). In direction *t* of Fig. 302 the plate is somewhat flexible and separates with a fibrous fracture, "bend cleavage."

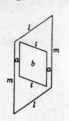

Fig. 302.  Gypsum Cleavage.

Gypsum has a hardness of only 2, and hence it is easily scratched by the finger nail. The specific gravity is 2.32. The luster is usually vitreous but may be silky or pearly on the side of the best cleavage. The selenite variety is clear and colorless; the massive kinds are generally snowy white, as in alabaster. Impurities may give it various shades of yellow, red, and brown.

**Composition.** Hydrous calcium sulfate, $CaSO_4 \cdot 2H_2O$. When a fragment is held in the flame it becomes opaque white and exfoliates, fusing to a globule. Abundant water is given off in the closed tube.

**Occurrence.** Gypsum is most commonly found as a sedimentary rock interstratified with limestones and shales and usually underlying beds of rock salt. Such beds result from the evaporation of salt water bringing about the crystallization and precipitation of gypsum and other dissolved salts. Many commercial deposits of gypsum are known in the United States. The principal producing states are New York, Michigan, Iowa, Texas, Nevada, and California. One of the most unusual occurrences of gypsum is in New Mexico at the White Sands National Monument. The gypsum, which has been deposited as the result of the evaporation of water in an enclosed basin, has been blown into great dunes. Here one can look for miles and see nothing but the rolling snow-white surface of the dunes.

Small crystals, such as those shown in Fig. 303, have formed in mud which permitted them to grow freely in all directions. Large fine crystals have come from Wayne County, Utah, and Naica, Mexico (Fig. 8, p. 19).

FIG. 303.  Gypsum Crystals, Ellsworth, Ohio.

## TUNGSTATES, MOLYBDATES, AND URANATES

### WOLFRAMITE, (Fe,Mn)WO$_4$

Wolframite is the chief ore of tungsten, a metal that has risen to great importance in our present civilization.  Its most familiar use is in the filaments of electric light bulbs where it is valuable because of its extremely high melting point, 3,350° centigrade. When the current is turned on, the filament becomes white hot without melting.  It will, however, oxidize when hot; our present lamps, therefore, are filled with an inert gas which insures the absence of oxygen.

The most important use of tungsten is as a steel-hardening metal, and tungsten is thus used in armor plate and metal-piercing projectiles.  High-speed cutting tools are made from tungsten steel and will retain their temper even when red hot.  Machines can thus be speeded up, and, as a result, more work can be turned out in a given time.  Tungsten carbide, harder than corundum, is used as an abrasive material for cutting glass and hard steel and, in some core drilling, replaces diamond.

**Habit.**  Wolframite is monoclinic.  Crystals are usually flattened parallel to the side pinacoid with a bladed habit.  It is found also in massive granular aggregates.

**Physical Properties.** Wolframite has perfect cleavage parallel to the side pinacoid. The hardness is 5–5½; the specific gravity is high, 7.0–7.5. The luster is submetallic; the color, brown to black. The streak is also brown to nearly black.

**Composition.** Wolframite is ferrous and manganous tungstate, $(Fe,Mn)WO_4$. It is an isomorphous mixture of iron tungstate, *ferberite*, and manganous tungstate, *huebnerite*. These two minerals are in general rare compared to wolframite. They have physical properties similar to those of wolframite and are difficult to distinguish from it without testing for the presence of iron or manganese.

**Occurrence.** Wolframite is usually found in quartz veins or pegmatite dikes, and more rarely in veins with sulfide minerals. About half of the world's supply of tungsten comes from China as wolframite. In the United States wolframite is found in the Black Hills of South Dakota, but only in small amounts. Ferberite from Colorado and huebnerite from Nevada, Colorado, and South Dakota are more important in the United States as ores of tungsten.

### Scheelite, $CaWO_4$

Scheelite is an ore of tungsten but of lesser importance than wolframite. It is named after K. W. Scheele, the discoverer of tungsten.

**Habit.** Scheelite is tetragonal. Crystals are usually simple dipyramids (Fig. 304). It is also found in granular aggregates.

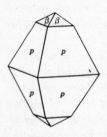

FIG. 304. Scheelite.

**Physical Properties.** Scheelite has cleavage (4 directions) parallel to the second-order dipyramid, the angles of which resemble closely those of the octahedron. The hardness is 4½–5; the specific gravity, 5.9–6.1, is unusually high for a mineral with nonmetallic luster. The luster is adamantine, and the color is white, yellow, green, or brown. Scheelite is almost unique among minerals in that most specimens of it will fluoresce a pale blue. This property, therefore, can be used in prospecting for the mineral or in gaining an idea of the amount present during mining.

**Composition.** Scheelite is calcium tungstate, $CaWO_4$. It is decomposed by boiling in hydrochloric acid, yielding a yellow residue of tungstic oxide.  If metallic tin or zinc is added and boiling is continued, the solution turns first blue and then brown.

**Occurrence.**  Scheelite is found in pegmatite dikes and in ore veins associated with cassiterite, molybdenite, and wolframite. It is thus found in the tin deposits of England and Australia. In the United States it is mined in Nevada, California, and, in smaller amounts, in Arizona, Utah, and Colorado.

## Wulfenite, $PbMO_4$

Wulfenite is an ore of molybdenum but of considerably less importance commercially than molybdenite.  It is, however, of interest to the mineral collector because of the great variety and beauty of many of its specimens.

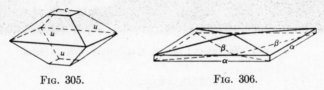

Fig. 305.          Fig. 306.

Wulfenite Crystals.

**Habit.**  Wulfenite is tetragonal, and crystals are usually square tabular in habit with a prominent base.  Others are as thin as a knife-edge with the flat table beveled by the faces of a low dipyramid (Fig. 306).

**Physical Properties.**  The hardness is 3; the specific gravity, 6.7–6.9.  The color is most commonly a bright orange-yellow to reddish yellow, but it may be green or brown.  The luster is resinous or adamantine.  The square habit, bright color, and high luster make it a most striking mineral.

**Composition.**  Wulfenite is lead molybdate, $PbMO_4$.  It fuses easily before the blowpipe and gives a lead globule when fused with sodium carbonate.

**Occurrence.**  Wulfenite is a secondary mineral found in the upper oxidized portion of lead veins associated with other secondary lead minerals, such as pyromorphite and vanadinite.  Beautiful crystals have been found in several places in western United States, particularly in Arizona and New Mexico.

Fig. 307.  Wulfenite Crystals, Las Cruces, New Mexico.

## URANINITE

If this book had been written before August, 1945, uraninite would have been mentioned as a rather rare mineral mined as an ore of uranium and of the very rare element radium.  Since that date, however, uraninite has taken on peculiar significance, for it was the chief raw material used in the construction of the atomic bomb.  The localities in which it had been mined became a matter of international concern, and a world-wide search for the mineral has been going on ever since.  Uranium will probably never again be used as an alloy in steel, a use to which it had formerly been put in a limited way.

Until the advent of the atomic bomb, interest in uraninite lay in the fact that it was the source of radium.  This element is and will continue to be in great demand.  Its extraction is a slow and costly process, for about 750 tons of ore must be mined to yield about 1 gram of radium.

**Habit.**  Uraninite is isometric, and crystals, though rare, show faces of the octahedron and the dodecahedron.  It is usually massive.

**Physical Properties.**  The hardness is 5½, and the specific gravity is 9–9.7.  This is unusually high but what one would expect, for uranium is one of the heaviest metals (G = 18.86). The luster is pitchlike, and because of this the mineral is commonly called *pitchblende*. The color is black, and the streak is brownish black.

**Composition.** Uraninite is essentially $UO_2$ and $UO_3$ but contains also small amounts of radium, thorium, and other rare elements. Small quantities of lead and helium are present as the result of radioactive disintegration.

**Occurrence.** With the great importance that uraninite has assumed in recent years as a source of uranium, many new localities have undoubtedly been discovered and exploited. These, however, remain unknown to the general public, and only the older localities can be mentioned. Two are important: one is in Katanga and the other in Canada on the shore of Great Bear Lake. Both localities have been worked for many years, chiefly for the small percentage of radium that could be extracted from the ore. Since 1940 mining has been accelerated, but this time for the uranium.

Small amounts of uraninite have been found at many localities associated both with pegmatites and with the ore minerals of silver, lead, and copper. Thus in the United States isolated crystals are found in the pegmatites of New England and North Carolina.

The presence of uranium is usually detected first by its bright-colored alteration products. Among the commonest of these are yellow *autunite* and *uranophane*, and orange *gummite*. *Carnotite*, also used as an ore of uranium, has been mined in Colorado and elsewhere.

## SILICATES

The silicates form the largest class of minerals. Relatively few of them are used as ores, but they are extremely important as rock-forming minerals. With only minor exceptions all the minerals of the igneous rocks are silicates, and thus they make up the bulk of the earth's crust. Of the minerals that are either common enough or important enough to be described in this book, nearly 40 per cent are silicates.

### QUARTZ, $SiO_2$

Quartz is the most common mineral, and in some of its varieties one of the most beautiful. It makes up most of the sand of the seashore; it occurs as a rock in the forms of sandstone and quartzite and is an important constituent of many other rocks, such as granite and gneiss.   It is a mineral which can usually be recognized by its form when crystallized, also by its hardness, conchoidal fracture, glassy luster, and infusibility.   There are so many

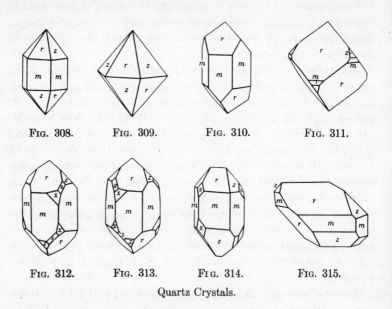

FIG. 308.          FIG. 309.          FIG. 310.          FIG. 311.

FIG. 312.          FIG. 313.          FIG. 314.          FIG. 315.

Quartz Crystals.

varieties, however, that it is only after long practice that one can be sure of always identifying it at once.

**Habit.**   Quartz is hexagonal, and the common habit of its crystals is a hexagonal prism showing horizontal striations terminated by six pyramidal faces (two rhombohedrons) each having the shape of an acute isosceles triangle (Fig. 308).   In some crystals the prism is not present and the shape is like that of Fig. 309, which appears to be a hexagonal dipyramid but is actually made up of two rhombohedrons.   It is not uncommon to find the faces of the rhombohedron lettered *r* much larger than those of the other rhombohedron, or they may be present alone.   (Figs. 310 and 311.)

On some crystals small modifying faces such as *x* and *s* in Figs. 312 and 313 may be present revealing the true symmetry of quartz. They appear rather complicated as, indeed, they are, and the study of the structure and crystallography is a matter for the skilled mineralogist. Nevertheless, even the beginner can learn to recognize the difference between the crystals represented by Figs. 312 and 313. Figure 312 is called a *right-handed* crystal and has the little *x* face to the *right* above the prism face *m*; the other is a *left-handed* crystal and has a similar face to the *left* above *m*.

Some crystals may be so poorly formed that the hexagonal nature is not immediately apparent. These, such as Fig. 315, can be oriented by the horizontal striations on the prism faces, and it should be remembered that in all cases the angles remain the same, in spite of the seeming irregularity in the form.

Though doubly terminated crystals, like those shown in Figs. 308 and 309, are found in some places, more commonly the prisms are attached at one end with only the other end free to develop. Some crystals may be slender and tapering. Not infrequently the crystals are so small that the forms can be distinguished only with a magnifying glass. A surface covered with many such tiny crystals is called *drusy*.

Twin crystals of quartz that show the re-entrant angles of most twins are not often found. Yet a careful study of almost any crystal will reveal that it is made up of two individuals in twin position interpenetrating each other irregularly. In order to prove such twinning it may be necessary to etch the crystal in hydrofluoric acid. One can then see that light is reflected differently from the etch pits formed on the two individuals. On some crystals natural etching has produced a difference in luster in various parts of a surface, thus outlining the twins.

The crystals illustrated have a three-fold symmetry axis and 3 two-fold symmetry axes but no symmetry center and are *low-temperature quartz*. When quartz is heated above 573° C, an internal rearrangement of the atoms forms *high-temperature quartz* with a different structure and higher symmetry.

Many massive varieties of quartz, some of which have been given names, will be mentioned in subsequent paragraphs.

**Physical Properties.** The hardness of quartz is 7; it cannot be scratched by a knife, but it easily scratches glass; the specific gravity of pure crystals is 2.66. The luster is vitreous in crystals, but in some massive kinds it may be greasy or waxy; the impure varieties, like jasper, are dull. The color varies widely; crystals are usually colorless but may be purple, yellow, and brown to nearly black; pink, green, and red kinds usually are not in crystals.

In the massive forms the color is often in bands or clouds, as described under the varieties.

**Composition.** Quartz is silicon dioxide, $SiO_2$. It is grouped with the silicates because its properties are more closely related to those minerals than to the oxides.

**Varieties.** Quartz is found in a great number of varieties, differing particularly in color and state of aggregation, and, as

Fig. 316.   Milky Quartz coated with Smaller Crystals, Ouary, Colorado.

many have been used for ornamental purposes, the varieties have received a number of distinct names. Those occurring in distinct crystals are named chiefly according to their color. They include:

*Rock crystal*, the clear colorless variety. If quite free from flaws, it is used for prisms for optical apparatus and cut into thin oriented plates for the control of radio frequencies. Most of this so-called optical quartz comes from Brazil, and during the war years 1942–1945 hundreds of tons of it were imported into the United States to be manufactured into radio oscillators.

*Smoky quartz*, having a smoky brown color, which in some crystals may be very dark. It is cut into ornaments, as in Switzerland, where it is found in beautiful specimens. Such crystals were undoubtedly at one time clear and have darkened by exposure to emanations from radioactive materials.  Some clear quartz can be artificially darkened by exposure to a strong x-ray beam.

*Amethyst*, a fine purple variety used for ornamental purposes.

*Citrine*, a light yellow-colored crystal sometimes called *false topaz*.  Some jewelers sell citrine under the name of *topaz*, and one should be careful not to confuse it with true topaz.

*Milky quartz*, milky white in color from the presence of small liquid inclusions (Fig. 316).  It is the common type of quartz found in veins and in pegmatites.  When two specimens of milky quartz are rubbed together, they luminesce.  The phenomenon is known as triboluminescence.

*Rose quartz*, a pale to deep pink variety.  It is used for ornamental purposes.  Rose quartz is usually massive, but small crystals have been found at Newry, Maine.

Although it is not found in crystals, *aventurine* should be mentioned here.  It is a variety spangled with scales of hematite, goethite, or mica.  *Cat's-eye* is a kind of quartz that gives when polished a peculiar effect of opalescence, due to the fibrous habit of the aggregate.  The name is given to other stones having similar effects.  It should not be confused with the highly prized cat's-eye of jewelry, which is a variety of chrysoberyl (p. 171).  *Tiger-eye* is a fibrous quartz of a yellow color and somewhat like cat's-eye in effect.  Its fibrous nature has been inherited from crocidolite from which it has formed by replacement.

The fine-grained or cryptocrystalline varieties of quartz are numerous and include many valued ornamental stones.  The most important of them will be described.  It should be remembered that, although these varieties differ physically from coarsely crystalline quartz, they are the same chemically and have the same internal structure.

*Chalcedony* is a variety having a waxy luster, either transparent or translucent, and varying in color from white to gray, blue, brown, red, and other shades.  It is often mammillary, botryoidal, or stalactitic.  Chalcedony is a general term, and specific names are given to some of the different colored varieties.

*Agate* is a variegated chalcedony with the colors arranged in delicate concentric bands frequently alternating with bands of opal, as in Figs. 317 and 318.  These bands often follow the irregular outline of the cavity in which the silica was deposited.  *Moss agate* is a kind of chalcedony containing brown or black moss-like or dendritic forms distributed rather thickly through the mass. These forms consist of some metallic oxide (manganese oxide is

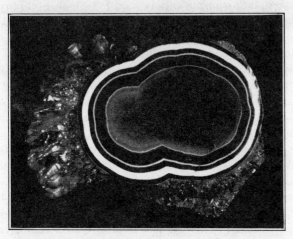

FIG. 317.   Agate, Brazil.

FIG. 318.   Agate, Brazil.

common) and have nothing more to do with vegetation than the frost figures on the windowpane in winter.

*Onyx*, like much agate, is made up of layers of different colored chalcedony and opal, but the banding is straight and the layers are in parallel planes. Alternating layers of white and black or white and brown are most common. Onyx is used for cameos, the head being cut from one layer and the background being formed by the other. Both agate and onyx are often artificially colored in order to make them more attractive for ornaments.

*Sardonyx* is like onyx but has layers of sard (carnelian) with other layers which are white or black. *Prase* is a translucent leek-green chalcedony; *chrysoprase* is an apple-green chalcedony. *Heliotrope*, or *bloodstone*, is a green chalcedony with small spots of red jasper scattered through it.

All the other varieties of cryptocrystalline quartz that are discussed are more or less impure; their luster is dull and in many cases they are almost opaque.

*Jasper*, an impure quartz, is usually red from inclusions of hematite. It may also be brown, yellow, or dark green. The green and red colors may be present in the same specimen irregularly distributed or arranged in bands.

*Flint* is nearly opaque with a dull luster and is usually gray, smoky brown, or brownish black. The exterior is often white from a mixture with lime or chalk in which it was originally embedded. It breaks with a conchoidal fracture, yielding sharp cutting edges, and hence was easily chipped by the American Indians into arrowheads and hatchets.

*Chert* is a compact silica rock resembling flint in most of its properties but is usually white or light gray in color. It frequently forms thin but extensive beds in limestone; flint is usually in isolated nodules. *Hornstone* is a name sometimes applied to chert.

*Silicified wood* consists largely of chalcedony or jasper which has replaced the woody structure of the tree. It may vary much in color and give beautiful effects on the polished surface, like the specimens from the "petrified forest" near Holbrook, Arizona. Some silicified wood has been formed by replacement by opal and properly belongs under that species.

**Occurrence.** Quartz is one of the essential constituents of granite, granodiorite, gneiss, mica schist, and many related rocks; it is the principal constituent of quartzite and sandstone. In these rocks it is found in small clear or milky grains that rarely show crystal outline. From the weathering and disintegration of such rocks quartz is set free to be washed into the streams and eventually into the ocean.   Thus it is the principal material of the pebbles of gravel beds and the sands of the seashore.   Such occurrences are familiar to all, but the mineralogist and crystallographer are more interested in the rarer localities where quartz is found well-crystallized or in the cryptocrystalline varieties.

Well-formed crystals may be found in veins with ore minerals or, as in Arkansas, in veins containing almost no other mineral.

Fig. 319.   Geode with Amethyst Crystals, Uruguay.

Some of the finest specimens have come from cavities in pegmatite and granitic rocks.   In cavities in traprock and in some limestones, chalcedony, agate, carnelian, etc., may be present filling all the available space; or they may only line the cavity and crystals of quartz may occupy the central part.   Embedded nodules or masses of fine-grained silica constitute the flint of the chalk formations, as in the chalk cliffs of Dover, England, and the chert of other limestones.

## TRIDYMITE AND CRISTOBALITE

Although the beginner cannot be expected to recognize tridymite and cristobalite, he should know what they are because they are of considerable interest to the mineralogist. Both minerals are high-temperature forms of silica and have the same composition as quartz, $SiO_2$. They are found, abundantly in some places, as constituents of fine-grained volcanic rocks. Cristobalite is stable only above 1470° C; tridymite, only above 870° C.

## OPAL, $SiO_2 \cdot nH_2O$

Opal is essentially silica but contains a few per cent of water; thus it does not have the same properties as quartz. Its chief interest lies in its use as a gem; stones are usually cut *en cabochon* to exhibit to best advantage the play of colors in the variety called *precious opal*. Another variety of opal, called *diatomaceous earth*, is white and chalky and one would never suspect that it is related to the beautiful gemstones. It is used as an abrasive, as a substance for filtering solutions, and as an insulation material.

**Habit.** Opal is one of the few minerals that are not crystalline and are called amorphous. It is found frequently in botryoidal or stalactitic masses. The variety diatomaceous earth resembles chalk in appearance.

**Physical Properties.** The hardness varies from 5 to 6, but opal can usually be scratched by a knife. The specific gravity varies from 1.9–2.2. It breaks with a smooth conchoidal fracture. The color may be white to yellow, red, brown, green, gray, and black. Several types of opal are recognized on the basis of their differing physical properties.

*Precious opal* is the most beautiful variety, much admired because of the delicate play of colors due to the optical effect of internal reflections. One kind of precious opal with a bright red flash of light is called *fire opal*. The beautiful opal found in Queensland, Australia, shows an iridescent blue like the effect of a peacock's feather.

*Common opal* does not exhibit the play of colors, and it varies widely in color and appearance and frequently has a resinous or waxy luster.

*Hyalite* is a clear glassy opal in globular or botryoidal masses that form crusts in rock cavities. In some specimens the glassy globules look like drops of gum.

*Wood opal* is silicified wood in which the mineral material is opal instead of quartz. Magnificent specimens of precious opal, faithfully preserving the external appearance of the branches they have replaced, are found in Humboldt County, Nevada. However, other wood opal breaks up into slender chalky-white splinters and shows none of the beauty of precious opal.

*Geyserite or siliceous sinter* is a type of opal deposited from hot springs, as by the geysers of Yellowstone National Park. It is usually white or gray and often has a pearly luster on the surface. It is soft and porous and frequently built up into concretionary forms of varied and beautiful appearance.

*Diatomaceous earth*, sometimes called *infusorial earth* or *diatomite*, is a kind of opal-silica consisting of the microscopic shells of the minute organisms called diatoms. As these shells sink from near the ocean's surface to the sea floor, they build up beds that may be of great thickness and extent.

**Composition.** The formula for opal can be written $SiO_2 \cdot nH_2O$. The $n$ indicates that it contains an indefinite amount of water. Opal, unlike quartz, is soluble in alkalies. Thus, after agate has been immersed in an alkaline solution for some time, the layers containing opal are attacked and dissolved leaving the layers of chalcedony unaffected.

**Occurrence.** Opal as a gemstone has been known for a long time. Most of the early material came from Hungary, but today much of the precious opal comes from Australia. In the United States it has been found in Nevada and Idaho. Deposits of diatomaceous earth are found in many places in the world; the most important in the United States are in California where it is found in beds over 4,500 feet thick.

## The Feldspars

The feldspars form the most important group of silicates, if not the most important of all the minerals. They are found as essential constituents of most crystalline rocks, such as granite, syenite, gabbro, basalt, gneiss, and many others, and thus make up a large percentage of the earth's crust. All of them are silicates of aluminum with potassium, sodium, and calcium, and rarely barium.

Although some feldspar is monoclinic and some triclinic, all occur in crystals that have a general resemblance to each other. All have cleavage in two directions, making angles of 90° or nearly 90° with each other. A careful examination will show that these two directions are unlike, that is, cleavage parallel to one face is easier than parallel to the other. The hardness of all the feldspars is about 6: they are not scratched by a knife. The specific gravity lies between 2.55 and 2.75, not far from that of quartz; the color is variable and may be white, pale yellow, reddish, greenish, or gray.

### ORTHOCLASE AND MICROCLINE, $KAlSi_3O_8$

Although orthoclase and microcline crystallize in different crystal systems, they have the same chemical composition, physical properties, and occurrence. Together they are known as *potash feldspar*. It is frequently difficult to distinguish between the two in the hand specimen, and thus in the past much feldspar has been called orthoclase that should be called microcline.

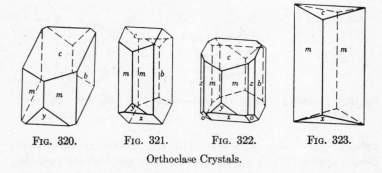

FIG. 320.     FIG. 321.     FIG. 322.     FIG. 323.

Orthoclase Crystals.

Potash feldspar is used extensively in the manufacture of porcelain, both in the body of the material and in the hard glaze with which much of it is finished. It is also a source of aluminum in the manufacture of glass. The green microcline known as *Amazon stone* is cut and polished and used for ornamental purposes.

**Habit.** Orthoclase is monoclinic; microcline is triclinic. The crystals illustrated in Figs. 320 and 321 are orthoclase, but Fig. 321 represents equally well a crystal of microcline. The near identity of crystals of the two species is due to twinning. Microcline, being triclinic, has no symmetry plane parallel to the side pinacoid, and this plane can thus be a twin plane on which the crystal can be repeatedly twinned. This is *albite twinning* and is always present in microcline giving it the apparent symmetry of a monoclinic crystal. Other types of twinning found in both species are: *Carlsbad* (Fig. 324), *Baveno* (Fig. 326), and

*Manebach* (Fig. 325). Figures 320 to 322 represent the common forms. The faces of the unit prism *m*, at angles of nearly 120° to each other, are short on some crystals (Fig. 320). The other most common forms are the basal pinacoid *c* and the side pinacoid *b*. There may also be other modifying planes.

**Physical Properties.** The name orthoclase is from the two Greek words meaning *erect* and *fracture*, referring to the existence of the two prominent cleavages at right angles to each other. One of these is parallel to the basal pinacoid, and the other is

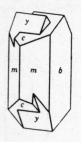 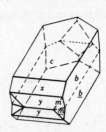 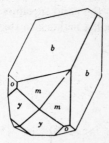

Fig. 324. Carlsbad Twin.    Fig. 325. Manebach Twin.    Fig. 326. Baveno Twin.

parallel to the side pinacoid. Although microcline receives its name from two Greek words meaning *little* and *inclined* (referring to the fact that the cleavages are slightly inclined from a right angle) they appear to be perpendicular because of the albite twinning. This is the outstanding physical property and is so characteristic of all the feldspars that it enables one to distinguish them quickly from other minerals even when they are in tiny grains in a rock.

The hardness is 6, and the specific gravity 2.57. The hardness is the best test by which feldspar can be distinguished from calcite, barite, and other minerals showing good cleavage. The luster is vitreous, though it may be pearly on the basal cleavage surface. The color is commonly white, reddish, or pale yellow. Some crystals of orthoclase are clear and colorless, and others, extremely rare, are a transparent yellow and are cut as gemstones. A green variety of microcline, *Amazon stone*, is also cut and used for ornamental purposes. (See frontispiece.)

The name *adularia* is given to a glassy variety of orthoclase that is usually found in pseudoörthorhombic crystals (Fig. 323). Some adularia gives a beautiful bluish opalescence, especially when polished, and is called *moonstone*. Some moonstone belongs to the albite species. *Sanidine* is a glassy orthoclase found in crystals embedded in various volcanic rocks.

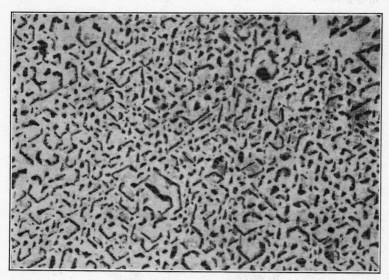

Fig. 327.  Graphic Granite, Bradbury Mountain, Maine
(quartz, dark; microcline, light).

**Composition.**  The formula for potash feldspar is $KAlSi_3O_8$. Some sodium may replace potassium, and in sanidine as much as 50 per cent of the potassium is replaced.  In the variety *hyalophane* barium replaces part of the potassium.  Chemical tests are difficult and unsatisfactory, and the beginner must rely on physical tests for identification.

**Occurrence.**  As has been mentioned, orthoclase and microcline are the common feldspar of granite, gneiss, and related rocks. They (particularly microcline) are present in large masses in granite pegmatites as found in the New England States, North Carolina, South Dakota, Colorado, and elsewhere.  In many of these pegmatites it is often possible to obtain feldspar in considerable quantity free from the associated quartz and mica.  It is then mined and used in making porcelain and glass.  Soda feldspar, albite, is commonly associated with the potash feldspar in pegmatites, as well as many interesting minerals such as tourmaline, beryl, apatite, amblygonite, spodumene, and many others. In some pegmatites microcline is intimately intergrown with quartz (Fig. 327). The name *graphic granite* is given to this intergrowth.

## ALBITE-ANORTHITE

Albite, $NaAlSi_3O_8$, and anorthite, $CaAl_2Si_2O_8$, are end members of an isomorphous series; that is, these two will mix together in all proportions so that crystals of any intermediate composition may be formed. This series is known collectively as the *plagioclase feldspars*, or, since one end member contains sodium and the other calcium, the *soda-lime feldspars*. Although there is a complete gradation from one end to the other, various species names have been given on a completely arbitrary basis to feldspar of intermediate composition. They are:

|  | Per cent Albite |
| --- | --- |
| Albite, $NaAlSi_3O_8$ | 100–90 |
| Oligoclase | 90–70 |
| Andesine | 70–50 |
| Labradorite | 50–30 |
| Bytownite | 30–10 |
| Anorthite, $CaAl_2Si_2O_8$ | 10–0 |

**Habit.** The plagioclase feldspars are triclinic. Crystals of albite and anorthite are more common than those of intermediate composition; all of them are usually small. Albite crystals are frequently flattened parallel to the side pinacoid and crowded together in parallel plates somewhat resembling barite. A snow-white albite in such platy aggregates is called *cleavelandite*.

Twinning is so common in the plagioclase feldspars that it is the rare crystal that shows none. *Albite twinning*, with the side pinacoid as the twin plane, is almost universal. This is a repeated twinning that is evidenced by parallel lines or striations seen on the basal pinacoid. In fact it is only by recognizing the presence of albite twinning that the elementary student can say that feldspar is triclinic rather than monoclinic. Carlsbad, Manebach, and Baveno twins (see p. 234) may also be present in the triclinic feldspars.

**Physical Properties.** The plagioclase feldspars have good cleavage parallel to both the base and the side pinacoid, and, although the angle between them is not 90°, the albite twinning makes it appear so. On the basal cleavage the evidence of this twinning can be seen by the fine lines which catch the light when the specimen is held so as to reflect it (Fig. 329).

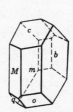

Fig. 328. Albite.     Fig. 329. Albite Twinning.

The hardness is 6; the specific gravity varies continuously from 2.62 in albite to 2.76 in anorthite. It may be colorless, white, or gray, and less frequently greenish, yellowish, or red. The luster is vitreous to pearly. A beautiful play of colors may be seen on some specimens of labradorite and andesine.

*Aventurine* or *sunstone* is a variety of oligoclase that contains inclusions of hematite that give a golden shimmer to the mineral. Some of the material known as *moonstone* is a variety of albite showing an opalescent play of colors. Some moonstone, as already noted, is orthoclase.

**Composition.** Sodium and calcium aluminum silicates; an isomorphous mixture of albite, $NaAlSi_3O_8$, and anorthite, $CaAl_2Si_2O_8$. Some potassium may be present particularly in the albite end of the series.

**Occurrence.** The plagioclase feldspars, like orthoclase and microcline, are rock-forming minerals and are even more widely distributed. They are present in most igneous rocks. Indeed, the classification of these rocks is based largely on the amount and kind of feldspar present. However, in general, it is true that the darker the rock the more calcium is present in the feldspar. In order to classify them correctly one must, for example, distinguish between oligoclase and labradorite, a task that must be left to the advanced mineralogist or petrographer.  All we can hope to do at this stage is to make a rough division on the basis of specific gravity, and even for this classification, a fragment larger than the average grain of an igneous rock is necessary.

In addition to its occurrence as a rock-forming mineral, albite is frequently found in pegmatite dikes.  It may be in small crystals or in larger masses replacing earlier microcline.  The platy variety, cleavelandite, is usually confined to pegmatites.

## The Feldspathoids

As has already been mentioned, the feldspars are to be found, usually abundantly, in most igneous rocks.  In a small group of igneous rocks, however, there is insufficient silica to combine with the alkalies and aluminum to form feldspars.  The rock minerals that form in their place with a lower percentage of silica are called *feldspathoids*.  The most important of these are leucite, nepheline, sodalite, and lazurite.

### Leucite, $KAlSi_2O_6$

**Habit.**  Leucite is pseudo-isometric.  It is nearly always found in trapezohedrons as shown in Fig. 330, and other forms are rare. It is said to be pseudo-isometric rather than isometric, for, if one examines a crystal under the polarizing microscope, he sees that

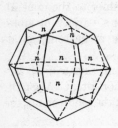

it is made up of many tiny orthorhombic crystals.  When these leucite crystals formed from hot molten magma they were truly isometric both in outward form and in internal structure.  However, on cooling below 500° C, an internal rearrangement took place but the external appearance remained unchanged.

**Physical Properties.**  The hardness is $5\frac{1}{2}$–6, and the specific gravity 2.45–2.50. (The specific gravity of analcime, the only

Fig. 330.  Leucite.

other mineral occurring in white trapezohedrons, is 2.27.)  The color is white to gray.  The name *leucite* comes from the Greek meaning *white*.

**Composition.**  Leucite is potassium aluminum silicate, $KAlSi_2O_6$. It contains the same elements as orthoclase feldspar but has only 55 per cent $SiO_2$ whereas orthoclase has 64.7 per cent $SiO_2$.

**Occurrence.** Leucite is a rare rock-forming mineral found in rocks with a deficiency of silica. Such rocks are limited in extent but are found in the Leucite Hills, Wyoming; and in the Highwood and Bear Paw Mountains, Montana. The lavas of Mount Vesuvius in Italy contain abundant crystals of leucite.

Since leucite is a rock-forming mineral, it is found embedded in a fine-grained matrix; analcime is found in isolated crystals or lining the cavities in dark-colored rocks.

## NEPHELINE, $(Na,K)(Al,Si)_2O_4$

**Habit.** Nepheline is hexagonal, but crystals are rare. It is usually massive, and is common as small grains in igneous rocks.

**Physical Properties.** Cleavage, parallel to the prism faces, is seen only in the larger crystalline units. The hardness is $5\frac{1}{2}$–6; the specific gravity is 2.55–2.65. In crystals the luster is vitreous, but when nepheline is massive the luster is greasy. For this reason the name *eleolite*, from the Greek word meaning *oil*, is sometimes given to the massive variety. The color is white to yellow in crystals, but the massive variety may be gray, greenish, or reddish

**Composition.** Sodium-potassium aluminum silicate, $(Na,K)(Al,Si)_2O_4$. It can be fused before the blowpipe, giving a strong yellow flame of sodium. It is soluble in hydrochloric acid and can thus be distinguished from feldspar.

**Occurrence.** Nepheline is a rock-forming mineral and is rarely found except in igneous rocks and pegmatite dikes associated with them. The granular rock, *nepheline syenite*, and the lava equivalent, *phonolite*, are the rocks in which it is found most abundantly. Masses of these rocks, however, are usually small. In the United States nepheline-bearing rocks are found near Magnet Cove, Arkansas, and Beemerville, New Jersey. Larger masses of igneous rock with coarse crystalline nepheline are found near Bancroft, Ontario.

## SODALITE, $Na_4Al_3Si_3O_{12}Cl$

**Habit.** Sodalite is isometric, and when crystals are found they are usually dodecahedrons. It is most commonly massive.

**Physical Properties.** Sodalite has dodecahedral cleavage. The hardness is $5\frac{1}{2}$–6; the specific gravity is 2.15–2.30. The color is usually blue but may also be white, gray, yellow, or red.

**Composition.**  Sodium aluminum silicate with chlorine, $Na_4Al_3$-$Si_3O_{12}Cl$.  It is fusible before the blowpipe, giving a strong yellow flame of sodium.  It is soluble in hydrochloric acid.

**Occurrence.**  Sodalite is a rather rare rock-forming mineral, associated with other feldspathoids, particularly nepheline.  It is found in crystals in the lavas of Vesuvius.  The massive blue variety has been found at Litchfield, Maine;  and in Canada in Ontario, Quebec, and British Columbia.

## LAZURITE, $Na_{4-5}Al_2Si_3O_{12}S$

Lazurite is the mineral name of the gem and ornamental stone *Lapis Lazuli*.  It has been known and highly prized by certain people for many centuries.  When ground to a powder it was used as the blue paint pigment, *ultramarine*.  This same pigment is now made synthetically.

**Habit.**  Lazurite is isometric, but crystals are rare.  It is usually massive and compact.

**Physical Properties.**  Lazurite has imperfect dodecahedral cleavage.  The hardness is 5–5½, and the specific gravity is 2.4.  The color is deep azure-blue to greenish blue.

**Composition.**  Sodium aluminum silicate with sulfur, $Na_{4-5}Al_2$-$Si_3O_{12}S$.  It fuses before the blowpipe, giving a strong yellow flame of sodium.  It is soluble in hydrochloric acid.

**Occurrence.**  Lazurite is a rare mineral occurring in limestones and usually associated with pyroxene and pyrite.  Most of the properties of lazurite are similar to those of sodalite, and the beginner may have some difficulty in distinguishing them.  However, the association of pyrite with lazurite is so common that it can be used as a determining criterion.  The best lapis lazuli has come from Afghanistan and is a mixture of lazurite with several other minerals.

## SCAPOLITE

**Habit.**  Scapolite is tetragonal, and crystals are usually prismatic showing prisms of both the first and second order and the dipyramid *r* (Fig. 331).

**Physical Properties.**  Scapolite has cleavage parallel to both the first- and second-order prisms;  there are thus four directions making angles of 135° with adjacent faces.  The hardness is 5–6, and the specific gravity is 2.65–2.74.  The luster is vitreous to pearly, and the color is commonly white to gray. There are also yellowish, reddish, greenish, and bluish varieties.

**Composition.** Scapolite is of varied composition; the name belongs more properly to a series rather than to a definite species. The name *wernerite* is given to the common scapolite. It is a complex sodium-calcium aluminum silicate with Cl, $CO_3$, and $SO_4$. *Marialite* is the end member of the series containing more sodium than calcium; *meionite*, the other end member, contains more calcium. Scapolite fuses easily before the blowpipe to a glass full of bubbles. It is partially decomposed by hydrochloric acid.

Fig. 331. Scapolite.

**Occurrence.** Scapolite characteristically is formed in limestones as the result of igneous intrusion and is associated with diopside, garnet, apatite, and sphene. It is found in large pink masses at Bolton, Massachusetts. Yellow crystals of gem quality have come from Madagascar.

## The Zeolites

The ZEOLITE FAMILY includes a number of beautiful minerals having a close relation to each other both in manner of occurrence and in chemical composition. They are all hydrous silicates; that is, they contain water that is given off when a fragment is heated in the closed tube. Like other hydrous silicates, they are of inferior hardness, chiefly $3\frac{1}{2}$–$5\frac{1}{2}$, and low specific gravity, ranging from 2.0 to 2.4. They are readily decomposed by hydrochloric acid. Many of them bubble up, or intumesce, when heated before the blowpipe, and this has given the name to the family from the Greek, *to boil.*

All the zeolites are said to be secondary minerals, which means that, unlike the feldspar, quartz, etc., which are part of the rock, they crystallized subsequent to the time of the formation of the rock in which they occur. They have been formed in most cases out of the materials of the feldspar or related minerals in the rock itself and hence occur usually in crevices, seams, or cavities, instead of in the solid mass.

All are silicates of aluminum with lime or soda or potash. Like the feldspars, they do not contain iron or magnesia; indeed, in the past they have been called *hydrous feldspars.*

The zeolites are commonly found associated with each other and with the minerals datolite, prehnite, apophyllite, pectolite, and calcite. The usual occurrence of this association is in cavities in the dark-colored "traprock" (see Fig. 332), such as forms the Palisades of the Hudson and is found in other places in New Jersey

FIG. 332.   Heulandite, Paterson, New Jersey.

as well as at various points in Connecticut and Massachusetts. Famous localities have been developed where railroad cuts or tunnels have been cut through ridges of this and similar rock (basalt), as at Bergen Hill, New Jersey, and similarly in British India.   Beautiful specimens come from Nova Scotia.

## HEULANDITE

**Habit.**   Heulandite is monoclinic and is characteristically found in crystals somewhat flattened on the side pinacoid and having a pseudoörthorhombic appearance.   (Fig. 333.)

**Physical Properties.** Heulandite has perfect cleavage parallel to the side pinacoid, and the luster, vitreous elsewhere, is pearly on this face. The hardness is 3½–4, and the specific gravity is 2.18–2.2. Heulandite is most commonly colorless or white, but it it may be red, gray, or brown.

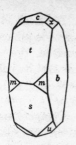

**Composition.** Heulandite is essentially a hydrous calcium aluminum silicate with a rather complicated formula. Some sodium and potassium may replace part of the calcium. Before the blowpipe it fuses to a white glass and gives water in the closed tube.

Fig. 333. Heulandite.

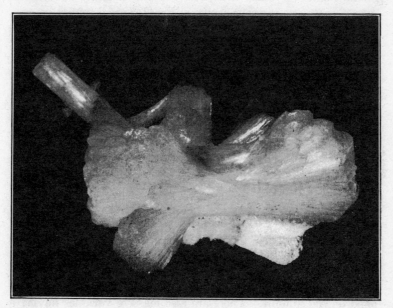

Fig. 334. Stilbite, Nova Scotia.

## Stilbite

**Habit.** Stilbite is monoclinic, and, although crystals are common, well-formed single individuals are rare. It is usually found in bundles of crystals, often looking like a sheaf of wheat tied tightly about the center (Fig. 334). This habit has given it the old name, *desmine*, from the Greek for *bundle*.

**Physical Properties.** Stilbite has perfect cleavage parallel to the side pinacoid; a beautiful pearly luster is seen on the cleavage surface, the side face of the bundles. This property gives the mineral its name from the Greek word meaning *luster*. The hardness is $3\frac{1}{2}$–4, and the specific gravity is 2.1–2.2. The color is usually white but may be yellow, red, or brown.

**Composition.** Stilbite is essentially a hydrous calcium-sodium aluminum silicate. Before the blowpipe it swells up and fuses to a white enamel. It yields water in the closed tube.

## CHABAZITE

**Habit.** Chabazite is rhombohedral and is usually found in rhombohedral crystals with nearly cubic angles. Penetration twins as shown in Fig. 335 are common.

**Physical Properties.** Chabazite has rhombohedral cleavage, which is much poorer than that shown in calcite. The hardness is 4–5. The specific gravity is 2.05–2.15. The luster is vitreous, and the color is white, yellow, and red.

**Composition.** Chabazite is essentially a hydrous calcium-sodium aluminum silicate with small amounts of potassium. Before the blowpipe it intumesces and fuses to a blebby, nearly opaque glass. It is decomposed by hydrochloric acid with the separation of silica.

FIG. 335. Chabazite.

## NATROLITE

**Habit.** Natrolite is monoclinic occurring usually in fine acicular, or needle-like, crystals. For this reason it is sometimes called the *needle zeolite*. These crystals are often arranged in radiating tufts lining cavities in the enclosing rock. When crystals are larger, they show that the form is nearly a square prism with a low pyramid on the summit. There are also massive varieties having a fibrous or fine-columnar radiated habit.

**Physical Properties.** The hardness is 5–$5\frac{1}{2}$; the specific gravity is 2.25. The luster is vitreous but may be pearly. Natrolite is commonly colorless or white but may be gray, yellow, or reddish.

**Composition.** Natrolite is a hydrous sodium aluminum silicate, $Na_2Al_2Si_3O_{10}\cdot2H_2O$. The composition is partially reflected in the name, which comes from the Latin word meaning *sodium*. Natrolite fuses to a clear transparent glass, coloring the flame

Fig. 336. Chabazite, West Paterson, New Jersey.

yellow. It is soluble in hydrochloric acid and gelatinizes upon evaporation.

## ANALCIME

**Habit.** Analcime is isometric and is usually found in trapezohedral crystals (Fig. 337) resembling one of the common forms of garnet. It less frequently occurs in cubes with three faces of the trapezohedron on each solid angle. In both form and color it resembles leucite, but its free-growing crystals, found in cavities in rocks, usually serve to distinguish it from leucite which as a rock-forming mineral is embedded in a fine-grained rock mass.

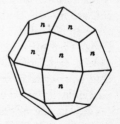

Fig. 337. Analcime.

**Physical Properties.** The hardness is $5-5\frac{1}{2}$; the specific gravity, 2.27; the luster, vitreous; the color, colorless or white.

**Composition.** Analcime is a hydrous aluminum silicate, $NaAlSi_2O_6 \cdot H_2O$. It fuses before the blowpipe to a colorless glass and gelatinizes with hydrochloric acid.

## CORDIERITE, $Mg_2Al_4Si_5O_{18}$

Cordierite, sometimes called *iolite* or *dichroite*, is a relatively rare mineral that is occasionally used as a gem. The name dichroite comes from a property shown by some specimens which present different colors when viewed in different directions.

**Habit.** Cordierite is orthorhombic, but crystals usually appear hexagonal because of twinning. It is most commonly in massive embedded grains.

**Physical Properties.** Cordierite has a poor cleavage parallel to the front pinacoid, but in many specimens cleavage is not observed. The hardness is $7-7\frac{1}{2}$, and the specific gravity is 2.6–2.66. It has a vitreous luster, and the color is various shades of blue. When viewed in transmitted light, some specimens show a deep blue color in one direction and a very pale blue color in another. Cordierite resembles quartz in many of its physical properties and can be distinguished from it with difficulty.

**Composition.** A silicate of magnesium and aluminum, $Mg_2Al_4Si_5O_{18}$. There are no easy diagnostic blowpipe or chemical tests for cordierite.

**Occurrence.** Cordierite is an uncommon mineral found in gneisses and schists and more rarely as an accessory mineral in granite. In Finland, Greenland, and Madagascar are notable localities. Gem material has come from Ceylon.

## KAOLINITE, $Al_2Si_2O_5(OH)_4$

Kaolinite is an important mineral for it makes up a large part of the soil mantle that covers such a high percentage of the rocks of the earth's crust. It is the chief constituent of kaolin or clay. Other clay minerals of lesser importance are *dickite, nacrite, halloysite,* and *beidellite.* All these minerals are of secondary origin; that is, they have been derived by alteration of aluminum silicates, particularly feldspar.

Clay is one of the most important natural substances used in industry. Common brick, tile, pottery, sanitary ware, and porcelain are a few of the clay products. The chief value of clay lies in the ease with which it can be molded into any desired shape; and then, when the clay is heated, part of the combined water is driven off, producing a hard, inert substance.

**Habit.** Kaolinite is monoclinic, but its closest approach to
crystals is thin, rhombic or hexagonal-shaped plates. It is usually
in fine-grained claylike masses.

**Physical Properties.** Kaolinite has perfect basal cleavage, but,
because of the minuteness of any individual particle, it is seldom
seen. The hardness is 2–2½; the specific gravity is 2.6. The
luster is earthy. When pure, kaolinite is white, but it is often
variously colored by impurities.

**Composition.** Kaolinite is a hydrous aluminum silicate,
$Al_2Si_2O_5(OH)_4$. Although it is a chemical compound corresponding to the formula given, it cannot be distinguished from the other
clay minerals without optical or x-ray tests.

**Occurrence.** As already mentioned, kaolinite is a widespread
mineral in the soils of the earth's surface, but as such it is relatively
impure. In certain places large masses of feldspar have altered
to pure kaolinite which can be mined and manufactured into white
porcelain or china. Some of the kaolinite formed by the weathering of feldspar remains in place to form a mantle of soil over the
rocks. Some of it is washed into the streams and may thus find
its way into the ocean or a lake to be deposited in layers that on
compaction are called clay. Such sedimentary clay deposits are
usually impure and thus cannot be used for high-grade china or
porcelain but are used in the manufacture of brick, tile, and other
ceramic products.

## Talc, $Mg_3Si_4O_{10}(OH)_2$

Talc is the softest of the minerals and is thus placed at the beginning of the scale of hardness. It is easily scratched by the finger
nail and has a slippery, soapy feel. For this reason the rock made
up mostly of talc is called *soapstone* or *steatite*.

Talc has many uses. In the powdered form it is used in the
manufacture of paint, paper, roofing material, and rubber as well
as in cosmetics in the form of face and talcum powder. Small
parts fashioned from talc and then fired in a furnace to drive off
the water become hard and durable and are used in electrical appliances. Slabs of the rock soapstone are used for laboratory table
tops, wash tubs, and sinks. Today less talc is used in this way
than formerly, for modern tubs, etc., made from steel with an
enamel coating are not only better looking but also more sanitary.

**Habit.** Talc is monoclinic. Crystals are rare, and those that are seen are usually thin plates with a rhombic or hexagonal outline. It is most commonly in foliated or compact masses.

**Physical Properties.**  Talc has perfect basal cleavage, but it can be observed in only certain types.  Plates of talc are flexible like those of mica, but, unlike mica, they are not elastic and when bent will not return to their initial position.  The hardness is 1, as stated above, and the specific gravity is 2.7–2.8.  The luster is pearly, especially in the foliated kinds, and the color in the finest of these is a beautiful sea-green.  There are also white foliated varieties.  The massive material may be white, dark gray, or green.

**Composition.**  Talc is a hydrous magnesium silicate, $Mg_3Si_4O_{10}$-$(OH)_2$.  It is an inert mineral and is unattacked by acids but yields water on intense heating in the closed tube.

**Occurrence.**  Talc is a mineral of secondary origin formed by the alteration of magnesium silicates.  It occurs most abundantly in the metamorphic rocks as soapstone associated with serpentine and chlorite.  There are many localities in eastern United States along the line of the Appalachian Mountains where talc has been mined or quarried.  It has also been mined in California.

## SERPENTINE,  $Mg_3Si_2O_5(OH)_4$

Serpentine is a remarkable mineral because it assumes a variety of massive forms (it is not known to occur in crystals).  The crystals of serpentine that are found are pseudomorphs, having been derived from some other mineral by chemical change.  Thus the magnesium-iron silicate, olivine, is often changed to serpentine, but the form of the olivine may be retained.

Serpentine assumes commercial importance because of the fine fibrous variety *chrysotile*, the most abundant type of asbestos.

Although a variety of amphibole is used to some extent as asbestos, chrysotile is more important and thus enters into the manufacture of incombustible fabrics.  Because of the flexibility of the fibers it is woven into cloth and gloves to be worn for tending hot furnaces and suits for fighting fire.  The shorter fibers are manufactured into products for insulating against heat and electricity.

A massive variety of green serpentine is called serpentine marble or *verdantique* marble and is used as a building stone.

**Habit.** Serpentine is monoclinic, but, as has been stated, crystals are unknown. The closest approach to a crystal is found in the fibrous variety, chrysotile. A platy type of serpentine is known as *antigorite*.

**Physical Properties.** The hardness varies from 2 to 5 but is usually about 4. The softer material is easily scratched and often has a smooth feel, sometimes almost greasy. The specific gravity

Fig. 338. Chrysotile Asbestos in Serpentine, Thetford, Quebec.

is 2.2 in fibrous varieties but may be as high as 2.65 when serpentine is massive. Light and dark shades of green may be present on the same specimen, giving a mottled appearance. The luster in the massive varieties is greasy or waxlike, but in the fibrous material it is silky.

**Composition.** A hydrous magnesium silicate, $Mg_3Si_2O_5(OH)_4$. Iron and nickel may be present in small amounts. When heated in the closed tube it yields considerable water and can thus be distinguished from varieties of amphibole that appear similar. It is decomposed by hydrochloric acid.

**Occurrence.** Since serpentine is formed by the alteration of such rock-forming minerals as olivine, pyroxene, and amphibole, it is a common and widely distributed mineral. Serpentine, as a rock name, is used for those rock masses made up largely of the variety antigorite. The serpentine asbestos is extensively mined at Thetford, Quebec, where it occurs in seams in a massive serpentine rock.   Some asbestos is found in Vermont and Arizona, but it is of little commercial importance.

## GARNIERITE

Garnierite is a mineral of secondary origin associated with serpentine and probably formed by the alteration of nickel-bearing olivine rocks.   It is a hydrous silicate of nickel and magnesium having a bright green color.   Garnierite appears to be amorphous and thus is found as incrustations or in massive form.   In certain places it has served as an ore of nickel and as such was mined in New Caledonia.   It has also been found at Riddle, Oregon.

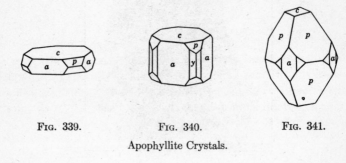

FIG. 339.              FIG. 340.              FIG. 341.

Apophyllite Crystals.

## APOPHYLLITE

**Habit.**   Apophyllite is tetragonal and occurs in square prismatic or pyramidal crystals but with a considerable variety in habit.   The crystals may be simple square prisms terminated by the basal pinacoid and may look like cubes.   Other crystals with these forms may also have the faces of a pyramid lying over the corners and thus appear similar to the combination of cube and octahedron.   It should be noted, however, that the angle made by the pyramid ($p$) and the base ($c$) is 119° 28′, not the same as that made by $p$ and the prism $a$ which is 128°.   Further, it is noticed on close examination that the faces $a$ have a vitreous luster and are vertically striated but the base $c$ has a pearly luster and is often dull.

**Physical Properties.**   Apophyllite has perfect cleavage parallel to the basal pinacoid, and hence the pearly luster is noticed on this face.   Elsewhere the luster is vitreous.   The hardness is 4½–5, and the specific gravity is 2.3–2.4.   The crystals are usually colorless or white but may show pale shades of green, yellow, or rose.

**Composition.** Apophyllite is a hydrous calcium potassium silicate containing a small amount of fluorine. It differs from the zeolites with which it is associated by containing no aluminum. Before the blowpipe it exfoliates, whitens, and fuses to a vesicular enamel. The name apophyllite refers to this property of exfoliating.

**Occurrence.** Apophyllite is found associated with zeolites, calcite, and datolite in cavities in traprocks. Beautiful crystals have been found in Bergen Hill and Paterson, New Jersey, and in Nova Scotia.

### Chlorite Group

The chlorite group is made up of several minerals that are so closely related that it is impossible for the beginner to distinguish between them. The principal members of the group clinochlore, penninite, and prochlorite are collectively described below under the heading of *chlorite*.

### CHLORITE

**Habit.** The chlorites are monoclinic and crystallize in tabular crystals with pseudohexagonal outlines; their habit is similar to that of the micas. Distinct crystals are rare, and chlorite is most commonly found in massive aggregates and finely disseminated particles.

**Physical Properties.** Chlorite has perfect basal cleavage. The thin cleavage plates can be distinguished from mica, for when they are bent they show no elasticity; that is, they will not return to the initial position as in the micas. The hardness is $2-2\frac{1}{2}$; the specific gravity, 2.6–2.9. The color is green of various shades. The name chlorite is from the Greek word for *green*, the same word which has given the name to the yellow-green gas chlorine, but there is no other relation between the chlorites and chlorine. Some chlorite is pale yellow, white, or rose-red.

**Composition.** The chlorites are hydrous iron-magnesium aluminum silicates. They fuse with difficulty and are unattacked by hydrochloric acid. At high temperature water is given off in the closed tube.

**Occurrence.** Chlorite is a common mineral usually forming as the result of the alteration of such minerals as pyroxenes, amphiboles, and garnet. Where rocks containing these minerals have undergone metamorphic changes, it is abundant; some schists are made up almost entirely of chlorite. Fine crystals of clinochlore have been found at the Tilly Foster Mine, New York.

## PREHNITE, $Ca_2Al_2Si_3O_{10}(OH)_2$

**Habit.** Prehnite is orthorhombic. It is seldom in distinct crystals but is usually in crystalline masses with a botryoidal or mammillary surface, or in groups of tabular crystals showing a series of faces in parallel position.

**Physical Properties.** The hardness is $6-6\frac{1}{2}$; the specific gravity, 2.8–2.9. The luster is vitreous, and, though the common color is green, it is sometimes white or gray.

**Composition.** Prehnite is a hydrous calcium aluminum silicate, $Ca_2Al_2Si_3O_{10}(OH)_2$. It fuses rather easily, yields a little water in the closed tube, and is slowly decomposed by hydrochloric acid.

**Occurrence.** Prehnite is associated with the zeolites and, like them, is a mineral of secondary origin. It is found lining cavities in basalts and related rocks. Beautiful specimens have come from Bergen Hill and Paterson, New Jersey.

### Mica Group

The mica group is made up of several closely related members that have many properties in common. Chief among these properties is the perfect cleavage, by means of which the micas may be split into leaves much thinner than a sheet of paper. In fact, it is difficult to set any low limit to the thickness of a cleavage plate. These leaves or sheets are usually very elastic and spring back when bent. Mica crystals are usually platy with an outline that is either rhombic with angles close to 120° and 60° or hexagonal with all the angles nearly 120°. The micas are complex aluminum silicates with potassium and hydroxyl and, in some varieties, sodium, lithium, magnesium, and iron.

Another group of minerals, the *brittle micas*, is similar to the micas in appearance. However, as the name implies, the folia are brittle and not elastic as the true micas. The most important brittle micas are *margarite*, *ottrelite*, and *chloritoid*.

## Muscovite, $KAl_3Si_3O_{10}(OH)_2$

Muscovite is variously called *white mica, common mica,* and *potash mica.* Plates of it, called "isinglass," are clear and transparent and have been used for many purposes. Because it is not affected by heat, muscovite is used for openings in stoves and furnaces. In the early days when glass was difficult to obtain, it was even used for the windows of houses, particularly in Russia, where muscovite was abundant (the old name for the country, *Muscovy,* was given to the mineral).

Today most of the uses of muscovite are based on its excellent properties as an electrical and heat insulator rather than on its transparency. Many of the small parts used for electrical insulation are punched from sheets of mica. One can see such a part in the top of each radio tube. Other parts are built up of thin sheets of mica cemented together and pressed into shape before the cement hardens. During World War II mica was such a critical material that many tons of muscovite were flown from India to the United States.

Much muscovite that is mined is too fine-grained to be used as "sheet mica" but is called "scrap mica." This is ground and used in the manufacture of wallpaper and roofing material. Mixed with oil it is used as a lubricant.

**Habit.** Muscovite is monoclinic, but crystals are rare. When occasionally seen they appear orthorhombic or hexagonal as shown in Fig. 342. The prisms usually taper sharply and are rough making the crystals appear dark. However, more light can pass through in this direction than perpendicular to the basal cleavage. The structure of the sheets, even when they have no regular shape, conforms to the pseudohexagonal outline; this is illustrated when a blunt point is held against a sheet and struck a blow with a hammer. A six-rayed star with branches intersecting at angles of 60° results. (See the center of Fig. 343.) Two of these branches are parallel to the prism faces, and the third is parallel to the side pinacoid. That the structure is pseudohexagonal is shown by inclusions of magnetite in muscovite that form a network along lines having the same directions. (Fig. 344.)

Muscovite most commonly occurs in scales or sheets without any regular form, its crystallization having been constrained by the surrounding minerals.

Besides occurring in distinct plates it assumes forms made up of minute scales, sometimes in compact aggregates, sometimes with a featherlike habit, as in *plumose mica*  Often the plates or scales are arranged in spherical aggregates.

**Physical Properties.**  As has already been mentioned muscovite has highly perfect cleavage parallel to the base permitting

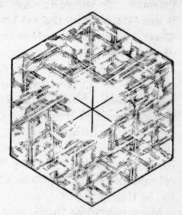

Fig. 342.  Muscovite Crystal.    Fig. 343.  Percussion Figure in Mica.

the mineral to be split into extremely thin sheets.  The hardness is 2–2½, and the specific gravity is 2.76–3.10.  The luster is vitreous to pearly.  In thin sheets muscovite is colorless and transparent, but in thicker blocks it may be dark brown.  Although smoky brown is the common color, it may be yellow, pink, or green.

**Composition.**  Muscovite is a silicate of aluminum and potassium having essentially the formula $KAl_3Si_3O_{10}(OH)_2$.  It yields a small amount of water when heated very hot in the closed tube. It is fused with difficulty and usually melts only on very thin edges.  It is not decomposed by acids.

**Occurrence.**  Muscovite is a common and abundant rock-forming mineral found in scales or plates in granite, gneiss, schist, and similar rocks.  In some pegmatite dikes, where the ordinary constituents of granite, feldspar, quartz, and mica are coarsely crystallized, large crystals of mica are found and may be mined with success.  In certain pegmatite dikes muscovite is found in immense sheets a yard or more across.  These dikes are very interesting to the mineralogist, for in them he looks not only for well-crystallized specimens of the three principal minerals named but also for many rarer minerals. Here are found tourmaline, beryl, apatite, garnet, and even rarer minerals, such as columbite, microlite, and uranium compounds.

Muscovite has been mined from pegmatite dikes in Maine, New Hampshire, Connecticut, North Carolina, South Dakota, and to a lesser extent in several other states. Much of the mica used in the United States is imported.

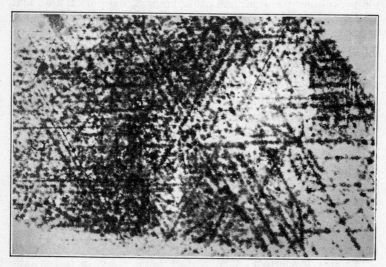

Fig. 344.   Oriented Magnetite Inclusions in Muscovite.

## BIOTITE,   $K(Mg,Fe)_3AlSi_3O_{10}(OH)_2$

Unlike muscovite, biotite has few uses and is of little commercial significance. However, when it has undergone a slight alteration to *vermiculite*, it is useful as a heat insulating material. When fresh biotite is heated there is no visible change, but when vermiculite is heated it swells up perpendicular to the cleavage to many times its former size. It appears to puff up leaving air space between the plates. It is an interesting experiment to heat several flakes of vermiculite and watch them apparently wiggle and squirm as they swell up.

**Habit.** Biotite is monoclinic. Although it may occur rarely in short prismatic crystals, it is usually found in scales with irregular boundaries or scaly aggregates.

**Physical Properties.** Like all the micas, the perfect basal cleavage is the outstanding physical property. The hardness is $2\frac{1}{2}$–3; the specific gravity, 2.8–3.2. The luster is bright and pearly on the cleavage surface, but vitreous on the edges. The color is dark green, brown, or black. The wide range in color corresponds to the variation in composition.

**Composition.** Biotite is a potassium magnesium-iron-aluminum silicate, essentially $K(Mg,Fe)_3AlSi_3O_{10}(OH)_2$. However, some kinds contain little iron, and other kinds contain much iron and almost no magnesium. The iron varieties have the darker color.

**Occurrence.** Biotite is one of the common rock-forming minerals and is found in small scales in almost all the igneous rocks. It is particularly common in rocks rich in feldspar, such as granite and syenite. In some pegmatite dikes it is found in large sheets but is less common than muscovite. It is also present in gneisses and schists, usually associated with muscovite.

## PHLOGOPITE,   $KMg_2Al_2Si_3O_{10}(OH)_2$

Phlogopite is a magnesium mica somewhat resembling biotite in color, but, because it lacks iron, it has desirable electrical properties. Sheets of it, therefore, are used for the same purposes as muscovite, chiefly as an electrical insulator.

**Habit.** Phlogopite is monoclinic and occurs in six-sided crystals which are often rough tapering prisms. It is found also in flakes and foliated masses.

**Physical Properties.** The perfect basal cleavage is the outstanding physical property. The hardness is $2\frac{1}{2}$–3, and the specific gravity 2.86. The color may be yellow, brown, green, or white and often has a copper-red appearance on the cleavage surface. The "star mica" of northern New York and Canada is a brown phlogopite which shows a fine six-rayed star when a point light source is viewed through it. (See Fig. 178, p. 85.) This is an example of asterism and is explained by the presence of minute rodlike crystals enclosed in the mica. These rods lie in positions parallel to the six-rayed star obtained by percussion (Fig. 343).

**Composition.** Phlogopite is a hydrous potassium-magnesium-aluminum silicate, $KMg_2Al_2Si_3O_{10}(OH)_2$. Unlike muscovite it is decomposed by boiling concentrated sulfuric acid.

**Occurrence.** Phlogopite is characteristically found in metamorphosed limestones and dolomites as isolated crystals disseminated through the rocks.

## LEPIDOLITE, $K_2Li_3Al_4Si_7O_{21}(OH,F)_3$

Lepidolite, frequently called *lithia mica*, is rare compared to the other micas and is found almost exclusively in pegmatites. In certain places it has been mined as a source of lithium for manufacture into various lithium compounds.

**Habit.** Lepidolite is monoclinic; crystals are rare. It is often found in masses that seem to have a close granular structure. Individual scales are sometimes large, and occasionally lepidolite is found in plates like other micas.

**Physical Properties.** Perfect basal cleavage is present as in the other micas but is not well shown in the granular compact varieties. The luster is pearly, and the color commonly pink to lilac, but it may also be a grayish white or yellow. The hardness is $2\frac{1}{2}$–4, and the specific gravity is 2.8–3.0.

**Composition.** Lepidolite is a potassium lithium aluminum fluosilicate; the composition can be expressed by the formula $K_2Li_3Al_4Si_7O_{21}(OH,F)_3$. It fuses easily before the blowpipe and, because of the presence of the lithium, it yields a red flame. In this manner it can be distinguished from muscovite.

**Occurrence.** As mentioned above, lepidolite is found in pegmatite dikes where it is associated with other lithium-bearing minerals. Chief among its associated minerals are the pink and the green varieties of tourmaline, spodumene, and amblygonite. It is found in various localities in Maine; Portland, Connecticut; the Black Hills, South Dakota; Pala, California.

## Amphibole Group

Included in the amphibole group of minerals are several species closely related to each other both chemically and structurally, although they crystallize in the orthorhombic, monoclinic, and triclinic systems. Furthermore members of this group resemble corresponding members of the pyroxene group (p. 260). It is therefore important to know how to distinguish members of one group from members of the other.

The advanced mineralogist can make use of the optical properties to distinguish the amphiboles from the pyroxenes, but the beginner must rely upon the physical properties, chiefly on the cleavage. In spite of the fact that the amphiboles crystallize in three different crystal systems, they all have similar prismatic cleavage. The intersection of these cleavage planes is at approximately 56° and 124° as shown in Fig. 345. The pyroxenes also have prismatic cleavage, but the prism faces intersect at nearly right angles (87° and 93°), as shown in Fig. 346. Thus, with a little practice, one can quickly tell whether the specimen in hand is an amphibole

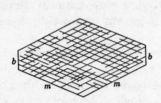

Fig. 345.  Amphibole Cleavage.     Fig. 346.  Pyroxene Cleavage.

or pyroxene. Although there are many exceptions, in general the amphiboles tend to be long and slender and the pyroxenes are short and stout.

Only two of the amphiboles, tremolite and hornblende, are common enough to warrant description here. Both of these are monoclinic.

## Tremolite, $Ca_2Mg_5Si_8O_{22}(OH)_2$

**Habit.** Tremolite is monoclinic and commonly in prismatic crystals. It is found also in bladed and columnar aggregates and in some places occurs in silky fibers that may be used as asbestos.

**Physical Properties.** Tremolite has perfect prismatic cleavage with the two directions intersecting at angles of 124° and 56° The hardness is 5–6, and the specific gravity is 3.0–3.3. The color varies from white to green; the darker green varieties are called *actinolite*. A closely compact tough variety is *nephrite* which furnishes much of the material known as *jade*. Jade is usually green but some specimens are mottled with nearly white to dark green patches.

**Composition.** The formula for pure tremolite is $Ca_2Mg_5Si_8O_{22}$-$(OH)_2$. However, iron is usually present replacing some of the magnesium and giving the mineral a green color. The higher the percentage of iron, the darker the mineral. It is called *actinolite* when iron is present in amounts greater than 2 per cent.

**Occurrence.** Tremolite is most commonly found in impure crystalline limestone and dolomite, where it is associated with other calcium silicates formed during the recrystallization of the rock. It is present also in certain talc schists. Actinolite occurs most abundantly in crystalline schists and in certain of them is the major constituent.

The finely fibrous asbestiform variety of tremolite has been found associated with metamorphic rocks in eastern United States. *Nephrite*, the tremolite which goes under the name of jade, has for centuries come from the Kuen Lun Mountains in southern Turkestan. Fine-quality jade comes from New Zealand and in recent years has been found near Lander, Wyoming. Another kind of jade is a pyroxene, *jadeite*. It resembles nephrite both in color and in the tough compact aggregates of fibers in which it occurs. It can be distinguished from nephrite by its easy fusibility.

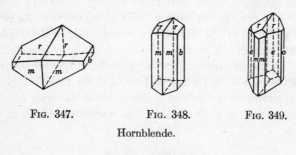

FIG. 347.          FIG. 348.          FIG. 349.

Hornblende.

## HORNBLENDE

**Habit.** Hornblende is monoclinic, and prismatic crystals terminated by a clinodome (*r*) are common (Figs. 347–349). It is found also in columnar and bladed aggregates.

**Physical Properties.** As in all amphiboles hornblende has perfect cleavage at angles of 56° and 124°. The hardness is 5–6, and the specific gravity 3.2. The luster is generally vitreous but is silky in fibrous varieties. The color is dark green to black, and, although it may appear opaque, it will transmit light on the thin edges.

**Composition.** Hornblende has such a complex chemical composition that it is difficult to write a formula for it. It is a silicate containing calcium, sodium, magnesium, iron, fluorine, and water. In addition it contains aluminum, an element not found in tremolite. Hornblende fuses with difficulty and yields water in the closed tube.

**Occurrence.** Hornblende is one of the common rock-forming minerals and is found in both igneous and metamorphic rocks. Many of the dark grains disseminated through light-colored syenite and granite are hornblende, and in other rocks this mineral may be present in amounts large enough to give them a dark appearance. Much hornblende found in igneous rocks has formed by the alteration of pyroxene. Many metamorphic rocks are rich in hornblende formed during the process of metamorphism at the expense of the pyroxenes.

## Pyroxene Group

The pyroxenes form one of the most important of the silicate groups. As stated on page 257 they are similar to the amphiboles in many respects. It is important enough to restate here that the only way for the beginner to distinguish between members of the two groups is by the cleavage. The cleavage of the amphiboles is at 124° and 56°, and the cleavage of the pyroxenes is at 93° and 87°. The pyroxene diopside is similar to tremolite both in its appearance and in its association, and the pyroxene augite is likewise similar to hornblende. Both these minerals are monoclinic, but there is another pyroxene, enstatite, which is orthorhombic.

### ENSTATITE, $MgSiO_3$

**Habit.** Enstatite is orthorhombic; all the other pyroxenes are monoclinic. Crystals are rare, and it is usually massive or fibrous.

**Physical Properties.** Enstatite has two good directions of cleavage at 87° and 93°. The hardness is $5\frac{1}{2}$, and the specific gravity is 3.2–3.5. The color of pure enstatite is grayish white, but when iron is present it may be yellowish, olive-green, or brown. *Bronzite* is a variety with a bronzelike luster.

**Composition.** Enstatite is magnesium silicate, $MgSiO_3$, when pure, but iron may be present replacing magnesium until the two are in equal amounts. The iron-rich variety is called *hypersthene* and is darker in color.

**Occurrence.** Enstatite is a common mineral in the dark igenous rocks and is found in pyroxenites, peridotites, gabbros, and both iron and stony meteorites. It is found in the United States at Edwards, New York; Texas; Pennsylvania; and Webster, North Carolina.

## DIOPSIDE, $CaMgSi_2O_6$

**Habit.** Diopside is monoclinic and is found in prismatic crystals showing a square or eight-sided cross section. It may also be granular.

**Physical Properties.** Diopside has imperfect prismatic cleavage at angles of 87° and 93°. A well-developed parting is present in some crystals parallel to the basal pinacoid. A less common parting parallel to the front pinacoid is characteristic of the variety *diallage*. The hardness is 5–6, and the specific gravity is 3.2–3.3. The color is white to light green and, in varieties rich in iron, may be deep green.

**Composition.** Diopside is a calcium magnesium silicate, $CaMgSi_2O_6$. The mineral *hedenbergite*, $CaFeSi_2O_6$, is isomorphous with diopside but with iron taking the place of magnesium; the two minerals form a complete isomorphous series.

**Occurrence.** Diopside is a mineral characteristic of contact metamorphic deposits where it is found in limestone associated with garnet, idocrase, tremolite, and sphene. It is also found in regionally metamorphosed rocks, and the variety diallage is present in some gabbros and peridotites. Pale green transparent crystals found at DeKalb Junction, New York, have been cut as gemstones.

## AUGITE, $Ca(Mg,Fe,Al)(Al,Si)_2O_6$

**Habit.** Augite is monoclinic, and crystals with a prismatic habit are common (Figs. 350–353). When the faces in the prism zone are the front pinacoid, side pinacoid, and prism, as in Fig. 350, they make angles of nearly 45° with each other.

**Physical Properties.** Prismatic cleavage yielding two directions at nearly right angles to each other is the most characteristic feature of augite, and by means of it one can distinguish it from hornblende. The hardness is 5–6; the specific gravity, 3.2–3.4. The color is dark green to black and in the average specimen appears opaque, for only on the thin edges will it transmit light.

**Composition.** The formula for augite is $Ca(Mg,Fe,Al)$-

$(Al,Si)_2O_6$.   It differs from the other pyroxenes in that it contains aluminum.

**Occurrence.**   Augite is the commonest of the pyroxenes and is present as one of the chief dark constituents of many igneous rocks.

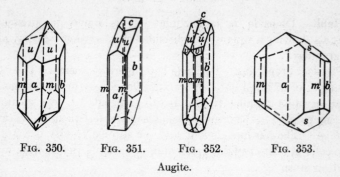

FIG. 350.        FIG. 351.        FIG. 352.        FIG. 353.

Augite.

It is particularly characteristic of the dark rocks, such as basalt, gabbro, and peridotite, and is less commonly found in the lighter-colored rocks, such as granite and syenite.

### SPODUMENE,  $LiAlSi_2O_6$

Spodumene is a rather rare mineral related to the pyroxenes but with a different type of origin and mineral association.   In certain places where it occurs abundantly it has been mined as a source of lithium.   Spodumene is of particular interest because of the great size of some of the crystals, some of which measured over 47 feet long.   Fine crystals of a lilac color, known as *kunzite*, and a deep green, known as *hiddenite*, are cut as gems.

**Habit.**   Spodumene is monoclinic and is frequently found in prismatic crystals with rough faces.

**Physical Properties.**   Spodumene has perfect prismatic cleavage at angles of 87° and 93°.   A parting, which in some specimens is more perfect than the cleavage, is usually present parallel to the front pinacoid.   The hardness is $6\frac{1}{2}$–7, and the specific gravity is 3.15–3.20.   The color most commonly is white to gray but, as already mentioned, may be lilac and green in the gem varieties.

**Composition.**   Spodumene is a lithium aluminum silicate, $LiAlSi_2O_6$.   Before the blowpipe it fuses to a clear glass coloring the flame red.

**Occurrence.** Spodumene is characteristically found in pegmatite dikes associated with other lithium-bearing minerals. The outstanding occurrence is at the Etta Mine in the Black Hills of South Dakota, where the gigantic crystals mentioned have been mined. It is also found in various pegmatites in New England. Kunzite is found in beautiful crystals at Pala, California, and in Madagascar. Hiddenite has been found at Stony Point, North Carolina.

## RHODONITE, MnSiO₃

Rhodonite is named from the Greek word for rose because of its beautiful rose-red color. It is not a common mineral but is found rather abundantly in some localities. One of the outstanding localities is in the Ural Mountains, where it has been mined and used as an ornamental stone. Here it is cut into thin slices for the veneering of table tops, etc.

**Habit.** Rhodonite is triclinic but closely related crystallographically to the pyroxenes. Crystals are usually flattened parallel to the basal pinacoid. Most commonly the mineral is in cleavable masses.

**Physical Properties.** Rhodonite has two directions of prismatic cleavage at angles of about 88° and 92°. The hardness is 5½–6, and the specific gravity ranges from 3.4 to 3.7. Rose-red is the most characteristic color, but it may also be pale pink or brown. The color of rhodonite is very similar to that of rhodochrosite, the manganese carbonate, but rhodonite can be distinguished from rhodochrosite by its greater hardness. Both these minerals may show a surface alteration of a black manganese oxide when exposed to atmospheric conditions for a short while.

**Composition.** The formula for rhodonite is MnSiO₃. *Fowlerite* is the name given to the zinc-bearing variety found at Franklin, New Jersey. Calcium also may be present replacing part of the manganese. Rhodonite is insoluble in acid, but fuses before the blowpipe to a black glass. When a small fragment is fused in a soda bead, a turquois blue color results indicating manganese.

**Occurrence.** As has been mentioned, the most important locality is in the Ural Mountains, but the best crystallized specimens have come from Franklin, New Jersey. A granular massive variety is found at Cummington, Massachusetts.

FIG. 354.   Pectolite, West Paterson, New Jersey.

## WOLLASTONITE

Wollastonite is found in crystalline limestone characteristically associated with diopside, tremolite, idocrase, and epidote. It is triclinic but crystals are uncommon; the usual occurrence is in coarsely cleavable aggregates. Cleavage parallel to the front and basal pinacoids yields fragments elongated parallel to the *b* crystallographic axis. The hardness is 5–5½; the specific gravity is 2.8–2.9. The luster is vitreous but may be silky in fibrous varieties; the color is white to gray.

Wollastonite is calcium silicate, $CaSiO_3$. It is decomposed by hydrochloric acid and, because of this fact, can be distinguished from tremolite which appears similar but is insoluble.

## PECTOLITE

Pectolite is not a common mineral but is often found with prehnite, datolite, and various of the zeolites. It is commonly in radiated or stellate aggregates or radiating acicular crystals

(Fig. 354). Distinct crystals are rare; these are triclinic elongated on the *b* crystal axis and show two cleavages parallel to the front and basal pinacoids. The hardness is 5; specific gravity, 2.7–2.8. The luster is silky to subvitreous; the color, white or grayish.

Pectolite is a hydrous calcium sodium silicate, $Ca_2NaSi_3O_8(OH)$. It yields water in the closed tube and fuses easily to a white enamel.

## CHRYSOCOLLA

Chrysocolla is a copper silicate formed by the oxidation of other copper minerals and is thus associated with malachite, azurite, cuprite, etc. It occurs in cryptocrystalline or massive aggregates of a blue-green or sky-blue color. Its hardness is 2–4; the specific gravity is 2.0–2.4. The formula is $CuSiO_3 \cdot 2H_2O$, but chrysocolla usually contains impurities. Chrysocolla resembles turquois but can be distinguished by its lower hardness.

## TOURMALINE

Tourmaline is one of the most attractive of the silicate minerals, and its varieties show a greater range of color than any other species. Fine-colored transparent crystals not only make superb mineral specimens but also have been cut and used as gemstones for many centuries.

Within recent years tourmaline has been used in certain scientific and industrial apparatus. This use, like one of the uses for quartz, is based on the property of piezoelectricity. Tourmaline is particularly suited for the construction of pressure gauges, and during World War II tourmaline gauges were used extensively in measuring blast pressures both in air and under water.

**Habit.** Tourmaline is rhombohedral. It is commonly in vertically striated prismatic crystals which show a triangular cross section. The three main faces are those of the trigonal prism, and the two smaller faces found at each of the points of the triangle are those of the second-order hexagonal prism. When crystals are doubly terminated, it is common to find a base at one end and a rhombohedron at the other. The commonest form is the obtuse rhombohedron (Fig. 355) which gives a terminal angle of about 133°. The different forms at the opposite ends of the crystal and the trigonal prism without parallel faces show that tourmaline belongs to a symmetry class lacking a center of symmetry. It should be remembered that only crystals without a center of symmetry show the property of piezoelectricity.

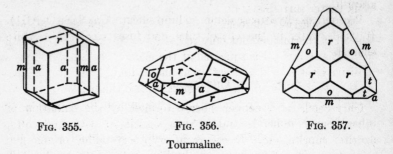

Fig. 355.       Fig. 356.       Fig. 357.

Tourmaline.

Although tourmaline commonly occurs in single crystals it is also found in radiating groups and in massive aggregates. The

Fig. 358.       Fig 359.

Tourmaline.

massive kinds usually show a columnar habit, the mass appearing as if made up of a bundle of parallel crystals. There are also varieties which are compact.

**Physical Properties.** Tourmaline has no distinct cleavage, but the fracture varies from uneven to conchoidal. It is brittle, and the fracture of a black mass often resembles that of a piece of coal. Its characteristic fracture enables one to distinguish it from aggregates of amphibole and epidote. The hardness is 7–7½, and the specific gravity varies from 3.0 to 3.25. The luster is vitreous, and the color, as already noted, is extremely varied. Black color is commonest, but there are also yellow, brown, blue, green, pink, and gray varieties; crystals are rarely white or colorless. Not uncommonly crystals differ in color in different parts. They may be pink at one end and green at the other, or there may be a pink center surrounded by a green exterior.

The property of piezoelectricity mentioned above has long been known in tourmaline, but only recently has any use been made of it. When a crystal is squeezed at the ends of the *c* crystal axis a positive electric charge is set up at one end and a negative charge at the other. The charge is proportional to the pressure exerted, and thus, with the proper recording equipment, it is possible to measure the pressure.

**Composition.** Tourmaline is a complex silicate of boron and aluminum with iron, magnesium, sodium, calcium, and lithium present in varying amounts. There is a rough correlation between color and composition. The common black tourmaline is rich in iron, the brown in magnesium, and the pink, green, and blue in lithium. The pink variety, called *rubellite*, is often associated with the lithia mica, lepidolite.

The brown magnesian varieties fuse rather easily to a blebby enamel, the iron kinds fuse with difficulty, and the lithia variety is infusible. If first powdered and then carefully mixed with three times its volume of powdered potassium bisulfate and its own volume of powdered fluorite and the mixture supported on a platinum wire in the blowpipe flame, it momentarily colors the flame green; this shows the presence of boron.

**Occurrence.** Tourmaline is a common mineral in gneisses, granites, and related rocks but is found especially well-developed in pegmatite dikes. In these dikes the black iron-rich variety may be found near the walls, having formed at the same time as the enclosing quartz and feldspar. In many pegmatite dikes only the black tourmaline is found, but near the center of some dikes fine crystals of the bright-colored gem material may be present in cavities. World-famous localities for gem tourmalines are in Brazil, Madagascar, and Russia; in the United States pegmatites in Maine, California, and Connecticut have yielded many beautiful gemstones.

The brown magnesian tourmaline is usually associated with crystalline limestone, and excellent crystals of this type have been found near Gouverneur, New York. Pierrepont, New York, is an outstanding locality for black crystals.

## BERYL, $Be_3Al_2Si_6O_{18}$

Beryl is the most common mineral containing the rare element beryllium and can thus be considered an ore of that interesting metal. Beryllium is much lighter than aluminum or magnesium and would thus have many uses if only it could be found in larger amounts. Since the supply is limited, beryllium is used only sparingly in alloys to which it gives many desirable properties.

However, it is not as an ore of beryllium but rather as a gem-stone that beryl has attracted attention through the centuries. The blue-green *aquamarine* and the deep green *emerald* are the gem

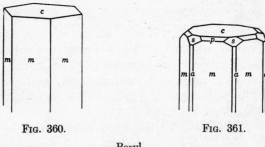

FIG. 360.                    FIG. 361.

Beryl.

varieties. Aquamarine is considered one of the semiprecious stones, but a fine-colored emerald may be of greater value than a diamond of comparable size.

**Habit.** Beryl is hexagonal and is one of the few species that is almost always in distinct and characteristic crystals which aid in identifying it. The common forms are the hexagonal prism and the base (Fig. 360); more rarely pyramid faces are present (Fig. 361). Some crystals may reach huge proportions, as at Albany, Maine, where crystals 3 feet in diameter and 18 feet long were found.

**Physical Properties.** Beryl has a poor cleavage parallel to the basal pinacoid, but it may not be observed in some specimens. The hardness is $7\frac{1}{2}$–8, or a little above that of quartz. The specific gravity is 2.75–2.80. The luster is vitreous, and in this respect beryl resembles quartz. The color is usually bluish green or light yellow, but it may be emerald-green, golden-yellow, pink, white, or colorless. When beryl is of gem quality various names are given to the different colored stones. The clear blue-green is *aquamarine*, the deep green *emerald*, pale pink to deep rose *morganite*, the clear yellow variety *golden beryl*.

**Composition.** Beryl is a beryllium aluminum silicate, $Be_3Al_2Si_6O_{18}$. In pink and colorless beryl small amounts of cesium are usually present replacing beryllium. The color of emerald is attributed to a minute amount of chromium. It is infusible before the blowpipe and is not attacked by acids.

**Occurrence.** Beryl is usually found in granitic rocks or in pegmatite dikes associated with them. The pegmatites of New England and North Carolina have afforded many beautiful specimens. The finest emeralds come from near Muso, Colombia, South America; others are found in the Ural Mountains, and some in North Carolina. Brazil is a source of both beautiful specimens of aquamarine and many tons of rough crystals used as a source of beryllium.

### HEMIMORPHITE

Hemimorphite, until recently called *calamine* in America, is a zinc silicate. It is a secondary mineral found in the oxidized portion of zinc deposits, having altered from sphalerite. It is orthorhombic but rarely found in isolated crystals. The common habit is in mammillary or botryoidal masses with crystalline surfaces (Fig. 363). Occasionally the surface is made up of flat tabular crystals projecting from the mass. The hardness is $4\frac{1}{2}$–5; the specific gravity, 3.4–3.5. It is usually white or slightly yellowish but may be tinged blue from a little copper. The luster is vitreous.

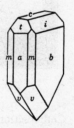

FIG. 362. Hemimorphite.

The composition is $Zn_4Si_2O_7(OH)_2 \cdot H_2O$ which gives 67.5 per cent ZnO. Thus where hemimorphite is abundant it is a valuable ore of zinc.

### OLIVINE, $(Mg,Fe)_2SiO_4$

Olivine, sometimes called *chrysolite*, is a common rock-forming mineral found most abundantly in the dark-colored rocks. The clear yellow-green variety is cut as a gemstone and goes under the name of *peridot*.

**Habit.** Olivine is orthorhombic. Crystals with good faces are rare and usually small. It is most commonly in small grains disseminated through certain igneous rocks.    Some rocks are made up almost exclusively of granular masses of olivine.

**Physical Properties.**    There is no cleavage but a pronounced conchoidal fracture.    It has a high hardness of 6.5–7; the specific gravity is 3.27–3.37.    The luster is vitreous, and the color yellow to olive-green and brown.

FIG. 363.    Hemimorphite, Sterling Hill, New Jersey.

**Composition.**    The formula for olivine, $(Mg,Fe)_2SiO_4$, represents an intermediate member in an isomorphous series, for the relative amounts of magnesium and iron vary between $Mg_2SiO_4$, *forsterite*, and $Fe_2SiO_4$, *fayalite*.    The most common olivines are richer in magnesium.    It is infusible before the blowpipe.

**Occurrence.**    Olivine is a rock-forming mineral found in some rocks only as scattered grains, but in others it may be the principal constituent.    It is most abundant in the dark rocks, such as gabbro, basalt, and peridotite.    In a certain rare type of rock known as *dunite*, it is the only important mineral present.    In addition to its presence in rocks, it is also a common mineral in some meteorites.    One celebrated meteorite found in Siberia in 1749 consists of a spongy mass of metallic iron containing bright yellow grains of olivine.    The transparent yellow-green gem material, peridot, is found on St. John's Island in the Red Sea

## WILLEMITE

Willemite, a valuable ore of zinc, is commonly found in bright yellow, apple-green, or brown masses and more rarely in small colorless or pale green hexagonal crystals. The variety known as *troostite* is sometimes found in large flesh-red crystals with a resinous luster. The hardness is $5\frac{1}{2}$, and the specific gravity is 3.9–4.2. It is a zinc silicate, $Zn_2SiO_4$. In troostite manganese is present replacing part of the zinc.

The most notable locality of willemite in the United States is at Franklin, New Jersey, where it is associated with calcite, franklinite, and zincite. However, unlike franklinite and zincite, which are peculiar to Franklin, willemite is found at other localities. One of the most interesting characteristics of willemite is its strong fluorescence, and some specimens show a marked phosphorescence as well (p. 86).

## CHONDRODITE

Chondrodite is a silicate of magnesium and iron containing fluorine. It occurs in yellow grains embedded in crystalline limestone, and also in deep red crystals as at Brewster, New York, where it is associated with magnetite. The hardness is $6-6\frac{1}{2}$; the specific gravity is 3.1–3.2. It is a member of the humite group of which *norbergite, humite,* and *clinohumite* are the other members. All these minerals have a similar appearance and composition, varying only in the percentages of the various similar chemical elements.

## GARNET

Garnet is the name given to a mineral group made up of several subspecies all of which have similar crystal habits, although their other properties vary widely.

Garnet is known chiefly through its use as an inexpensive red gemstone. However, the green andradite garnet, known as *demantoid,* is highly prized and is cut into beautiful gems. Garnet also has an industrial use as an abrasive, and the crushed mineral is used for sawing and grinding stones. It is also made into garnet paper that is preferred to sandpaper for certain purposes.

**Habit.** Garnet is isometric and almost always in distinct crystals, and, since the crystals are commonly isolated and disseminated through the rock, it is not difficult to recognize them.

There are, however, massive kinds which require some skill for their identification.

The common crystal forms are the dodecahedron (Fig. 365) and the trapezohedron (Fig. 366), or a combination of these two

FIG. 364.   Garnet Crystals in Mica Schist.

(Fig. 367).   More rarely the dodecahedron may have planes of the hexoctahedron (Fig. 368).   It should be remembered that the dodecahedron has angles of 120° between two adjacent faces and

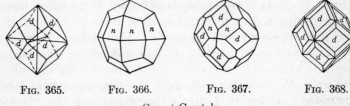

FIG. 365.        FIG. 366.        FIG. 367.        FIG. 368.

Garnet Crystals.

that these faces themselves are diamond-shaped with plane angles of 60° and 120°.

**Physical Properties.**   The hardness is $6\frac{1}{2}$–$7\frac{1}{2}$, and the specific gravity varies from 3.5 to 4.3.   The luster is vitreous, and the color, while most commonly red, varies from colorless to yellow, brown, black, and green.

**Composition.**   Although the compositions of the different garnets vary greatly, they can be expressed by the general formula $R_3''R_2'''(SiO_4)_3$.   In this formula $R''$ may be calcium, magnesium, iron, or manganese, and $R'''$ may be aluminum, iron, titanium, or chromium.   The subspecies are listed below with their chemical formulas and specific gravities, but it should be pointed out that between these there are also many intermediate varieties.

| Name | Composition | Specific Gravity |
|---|---|---|
| Pyrope | $Mg_3Al_2(SiO_4)_3$ | 3.51 |
| Almandite | $Fe_3Al_2(SiO_4)_3$ | 4.25 |
| Spessartite | $Mn_3Al_2(SiO_4)_3$ | 4.18 |
| Grossularite | $Ca_3Al_2(SiO_4)_3$ | 3.53 |
| Andradite | $Ca_3Fe_2(SiO_4)_3$ | 3.75 |
| Uvarovite | $Ca_3Cr_2(SiO_4)_3$ | 3.45 |

*Pyrope*, or magnesia garnet, often has a deep red color. When perfectly clear, it is used as a gem and called *precious garnet*.

*Almandite*, or almandine garnet, includes a large part of the common garnet; the rest of the common garnet is andradite. It is a silicate of iron and aluminum and is ordinarily red but sometimes black. When clear, it is cut into gems and like pyrope is called precious garnet.

*Spessartite*, or manganese garnet, is comparatively rare. It has a brownish red or hyacinth-red color. One type found in Virginia is perfectly transparent with a peculiar red shade and is used as a gem.

*Grossularite*, the calcium aluminum garnet, may be colorless, white, pale yellow, green, brownish yellow, or cinnamon-brown, and occasionally rose-red. The commonest kind is brownish red, and is called *cinnamon stone* or *essonite*. The original grossularite was green and thus received its name from the botanical name for the gooseberry.

*Andradite* is a calcium iron silicate and makes up much of the common garnet. The color is various shades of yellow, green, brown to black. *Topazolite* is a topaz-yellow variety; *melanite* is black; *demantoid*, the green gem variety with a brilliant luster is found in the Ural Mountains.

*Uvarovite*, or chrome garnet, is a rare kind of a fine emerald-green color. The finest specimens come from the Ural Mountains where it is associated with chromite.

Uvarovite is infusible, but all the other garnets fuse easily; if the fused mass is magnetic, it indicates that much iron is present.

**Occurrence.** Garnet is a common mineral found as an accessory constituent in granitic rocks but more characteristically and abundantly associated with metamorphic rocks. Almandite or andradite are the types usually found in mica schists; grossularite, in crystalline limestone; pyrope, in peridotites or serpentine; uvarovite, in cracks in chromite. As mentioned in the first pages of this book, garnet is found in rounded grains in streams and may make up a high proportion of the black and red sand of the seashore.

## IDOCRASE

Idocrase has until recently been called *vesuvianite* in America because the rocks of that famous volcano have furnished some of the finest specimens. Idocrase is of little economic importance, but a compact green variety found in California and known as *californite* has been used as a gem.

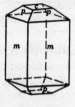
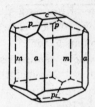

FIG. 369.                    FIG. 370.

Idocrase.

**Habit.** Idocrase is tetragonal and occurs in crystals of varied habit, some highly complex. A simple crystal is shown in Fig. 370 with a dipyramid, basal pinacoid, and the first- and second-order tetragonal prisms. It is found also in massive form and in striated columnar aggregates.

**Physical Properties.** The hardness is $6\frac{1}{2}$, and the specific gravity is 3.35–3.45. The luster is vitreous, inclining to resinous, and the color is commonly brown to green. It may also be sulfur-yellow and bright blue. Transparent crystals have been found but are rare.

**Composition.** The composition of idocrase is complex but can be expressed by the formula $Ca_{10}Al_4(Mg,Fe)_2Si_9O_{34}(OH)_4$. Boron or fluorine is present in some varieties, and beryllium has also been reported. Before the blowpipe it fuses with intumescence.

**Occurrence.** Idocrase is most commonly found in crystalline limestones, but also in serpentine and chlorite schists. It is frequently associated with grossularite garnet, diopside, epidote, and sphene. It resembles some brown garnet and brown tourmaline but is more fusible and in crystals is easily distinguishable.

## Epidote, $Ca_2(Al,Fe)_3(SiO_4)_3(OH)$

**Habit.** Epidote is monoclinic and frequently found in crystals elongated parallel to the *b* crystallographic axis. (Figs. 371 and 372.) If the long dimension is placed in a vertical position, it is difficult to find the symmetry plane which shows the monoclinic character. Epidote also occurs in fibrous or columnar aggregates and others that are compact and granular.

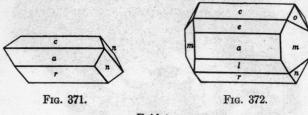

Fig. 371.     Fig. 372.

Epidote.

**Physical Properties.** Epidote has a perfect cleavage parallel to the base and an imperfect cleavage parallel to the front pinacoid. Crystals, therefore, break into fragments with the long dimension parallel to the *b* crystal axis. The hardness is 6–7, and the specific gravity is 3.35–3.45. The luster is vitreous, and the color is commonly green. The peculiar yellow-green or pistachio-green of ordinary epidote is so characteristic that in the majority of specimens color alone distinguishes it from other minerals that it may resemble. Well-formed crystals may be black, but, if they have been bruised or cracked, the green color is usually visible.

**Composition.** Epidote is a silicate of aluminum, iron, and calcium, $Ca_2(Al,Fe)_3(SiO_4)_3(OH)$. The ratio of iron to aluminum varies widely. It fuses with intumescence rather easily and, because of the iron it contains, gives a magnetic globule.

**Occurrence.** Epidote is a common mineral in many crystalline rocks, such as gneiss, mica schist, and amphibole schist. It also is found in contact metamorphic deposits in limestone. The finest crystals have come from Knappenwand, Austria, and from the Prince of Wales Island, Alaska. (See Fig. 373.) In the United States outstanding localities are Haddam, Connecticut and Riverside, California.

*Allanite* and *clinozoisite* are closely related to epidote and belong to the epidote group of minerals. Allanite is a black radioactive mineral containing cerium and other rare elements. Clinozoisite is a white to gray, greenish, or pale red mineral, like epidote in composition, except that it contains almost no iron.

Fig. 373. Epidote with Quartz, Prince of Wales Island, Alaska.

## Zircon, ZrSiO₄

Zircon is of interest because of the beautiful gems that have been cut from it. The common zircon gem is colorless, but red crystals called *hyacinth* are particularly prized. The blue gems sold under the name of "starlite" are zircons that have been colored artificially by heat treatment. Because of its high index of refraction, a well-cut colorless zircon resembles the diamond.

Zircon is the chief zirconium mineral and can be considered an ore of that rare metal, which is used in certain alloys. The rare mineral *baddeleyite*, found in Brazil, is also an ore of zirconium and a source of zirconium oxide, which is used in making refractory bricks.

**Habit.** Zircon is tetragonal and is almost always found in small prismatic crystals. The common forms are the first-order tetragonal prism and pyramid (Fig. 374). More highly modified crystals also occasionally occur (Figs. 375, 376).

**Physical Properties.** The hardness is 7½, and the specific gravity is 4.7. The luster is brilliant adamantine, and the color varies from colorless through various shades of reddish brown and yellow.

**Composition.** Zircon is zirconium silicate, $ZrSiO_4$. It is infusible and is not attacked by acids.

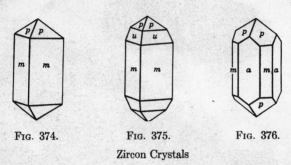

FIG. 374.  FIG. 375.  FIG. 376.

Zircon Crystals

**Occurrence.** Zircon is a common accessory mineral in all types of igneous rocks but is particularly abundant in granites and syenites. It usually occurs in these rocks in well-formed crystals, so tiny that it is difficult to see them well without a microscope. Larger crystals are sometimes found in crystalline limestone and in various other crystalline rocks. Fine specimens are obtained in New York, North Carolina, and Colorado, and very large ones in Renfrew County, Ontario, Canada (Fig. 377). Because zircon is both mechanically strong and chemically inert, it resists the normal processes of weathering, and in some regions it occurs in the rock so abundantly that when the rock is weathered away it is left unaltered in the soil. In other places these zircon grains are washed into the streams and, because of their high specific gravity, can be washed from the gravels after the manner of the placer gold miner. Concentrations of zircon in beach sand have been found in Florida, India, and Brazil.

## DATOLITE, $CaBSiO_4$

**Habit.** Datolite is monoclinic and commonly occurs in crystals that are complex and difficult to decipher even for the experienced crystallographer. The fact that many crystals are nearly equidimensional in the three axial directions (Fig. 378) adds to the difficulty of orientation.

FIG. 377.   Zircon Crystal in Calcite, Renfrew County, Ontario.

A kind of datolite that occurs in massive aggregates resembles porcelain.   Some of this type found in the Lake Superior copper-mining region is colored pink by finely disseminated copper enclosed within it.

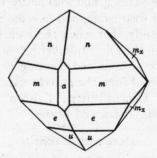

FIG. 378.   Datolite.

**Physical Properties.**   The hardness is 5–5½, and the specific gravity 2.8–3.0.   Crystals are usually clear and glassy with a faint greenish tinge.

**Composition.** Datolite is a calcium borosilicate, $CaBSiO_4$. It yields a little water when heated in the closed tube and gives the green flame of boron when held in the blowpipe flame, at the same time fusing easily.

**Occurrence.** Datolite is a mineral of secondary origin found in cavities and cracks in basaltic lavas and related rocks. It has the same origin as and is frequently associated with the zeolites, prehnite, and apophyllite. In the United States well-formed crystals of datolite are found in Massachusetts, Connecticut, and New Jersey.

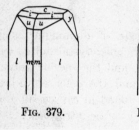
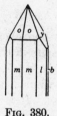
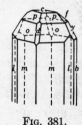

| Fig. 379. | Fig. 380. | Fig. 381. |
|:---:|:---:|:---:|

Topaz.

## Topaz, $Al_2SiO_4(F,OH)_2$

Topaz is highly prized as a gem mineral, and, although gem quality material occurs in several colors, the wine-yellow stones from Brazil are of most value and are called *precious topaz*. The pink color of some gem topaz is usually obtained by gently heating the dark yellow stones.

**Habit.** Topaz is orthorhombic and prismatic crystals such as shown in Figs. 379 to 381 are common. Frequently beautiful etch figures are found on certain faces but lacking on others; thus in a striking manner the difference in crystal structure parallel to different faces is shown. Although it is usually in crystals, it occurs also in massive crystalline aggregates. Most crystals are small, but some clear and well-formed crystals weighing as much as 600 pounds have been found in Brazil.

**Physical Properties.** Topaz has perfect basal cleavage. The hardness is 8, greater than that of any other common mineral with the exception of corundum. The specific gravity is 3.4–3.6. The luster is vitreous, and the color varies from colorless to white, straw-yellow, wine-yellow, pink, bluish, and greenish.

**Composition.** Topaz is an aluminum fluosilicate, $Al_2SiO_4$-$(F,OH)_2$. It is infusible and insoluble.

**Occurrence.** Topaz is associated with granites and is frequently found in related pegmatite dikes. Minerals commonly occurring with it are: tourmaline, cassiterite, beryl, quartz, and feldspar. It is found also in cavities in rhyolite lavas as in the Thomas Range, Utah. Fine wine-yellow and pale blue crystals have been found in Siberia. Other outstanding localities are Brazil (extremely large crystals), Japan, and Mexico. In the United States well-formed crystals have been found in California, Colorado, and Maine.

### AXINITE

Axinite is a rather rare borosilicate of aluminum, calcium, manganese, and iron. It occurs in triclinic crystals, often with a sharp edge. The hardness is $6\frac{1}{2}$–7, and the specific gravity is 3.27–3.35. The luster is vitreous, and the color clove-brown, violet, gray, green, or yellow. It is found in cavities in granite and in the contact zones surrounding granitic intrusions.

FIG. 382. Chiastolite.

### ANDALUSITE, $Al_2SiO_5$

Andalusite is named from the locality where it was identified, Andalusia in Spain. The most interesting variety is that called *chiastolite*. In this, parts of the crystals are white, and others contain carbonaceous impurities and are black. These inclusions are arranged in a regular manner to give a cruciform design in cross section (Fig. 382). The form on the cross section is a little like the Greek letter $\chi$, and this resemblance has given it the name chiastolite.

Andalusite has been mined in large quantities in California for use in the manufacture of highly refractory porcelain, such as that used in spark plugs.

The compound $Al_2SiO_5$ is polymorphous; that is, there is more than one substance with this composition; sillimanite and kyanite have the same chemical formula as andalusite.

**Habit.** Andalusite is orthorhombic and is usually in coarse prismatic crystals. The crystals are usually embedded and may be seen only when the surrounding rock is weathered away.

**Physical Properties.** The hardness is $7\frac{1}{2}$, and the specific gravity is 3.16–3.20. The luster is vitreous, and the color varies from white or gray to pink, brown, or green.

**Composition.** Andalusite is an aluminum silicate, $Al_2SiO_5$. It is infusible and insoluble.

**Occurrence.** Andalusite is formed most commonly by the regional metamorphism of shales into crystalline schists. The crystals are surrounded by fine-grained micaceous rock that weathers easily and leaves the andalusite protruding. Much andalusite, such as that found in New England, has been altered to fine-grained muscovite and, therefore, does not have the high hardness of the original mineral.

In the United States it is found in California, Maine, New Hampshire, and Pennsylvania. Chiastolite is found at Lancaster and Sterling, Massachusetts.

## SILLIMANITE

Sillimanite occurs in orthorhombic prisms that are usually long and slender. Sometimes they are aggregated into parallel groups that may be fibrous. Hence the name *fibrolite* has been used for the mineral. Sillimanite has a very perfect cleavage parallel to the length of the prisms. The hardness is 6–7, and the specific gravity is 3.23. The luster is vitreous; the color, pale brown to gray and green. The chemical composition is the same as that of andalusite. Sillimanite is a rare mineral found usually in gneisses and schists.

## KYANITE, $Al_2SiO_5$

Kyanite is named from the Greek word for blue because of its blue color. When it is transparent and of a rich color, fine gems can be cut from it. Like andalusite, it is used in the manufacture of high-quality refractories.

**Habit.** Kyanite is triclinic and is usually in long tabular crystals or in bladed aggregates.

**Physical Properties.** Kyanite has perfect front-pinacoid cleavage which is parallel to the length of the crystals. It shows better than any other common mineral the variation in hardness with crystallographic direction, for parallel to the length of the crystals it is 5, but across the crystals it is 7. The specific gravity is 3.56–3.66. The luster is vitreous to pearly. Some bladed crystals may be a fine blue throughout, but others show a central blue strip between paler or even colorless sides. It may also be white, gray, or green.

**Composition.** Aluminum silicate, like andalusite, $Al_2SiO_5$. It is infusible and insoluble.

Fig. 383. Kyanite and Staurolite (dark) in Mica Schist, St. Gothard, Switzerland.

**Occurrence.** Kyanite is found in gneisses and schists, usually as an accessory mineral, but in some places it makes up a high percentage of the rock. Fine specimens have come from Switzerland. In North Carolina and northern Georgia it is mined for use as a refractory.

### STAUROLITE, $FeAl_5Si_2O_{12}(OH)$

Staurolite is named from the Greek words meaning *cross* and *stone* because of the remarkable crosses formed by its twinned crystals. (Figs. 384, 385.) It is not unusual in a Christian country for superstition to attach great importance to these "cross stones" or "fairy stones." As a result the right-angle twins have been made up into amulets and sold to tourists. In certain places the demand has exceeded the supply, and the natives carve crosses out of fine-grained mica schist to sell to the unsuspecting customer.

**Habit.** Staurolite is orthorhombic. Single crystals are not rare, but it is more common to find twins with two crystals crossing each other nearly at right angles (Fig. 384) or at an angle of nearly 60° (Fig. 385). More complex twins, of the same interpenetration type, occur also where three or even four crystals are grouped together.

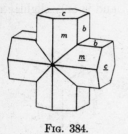
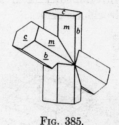

Fig. 384.    Fig. 385.

Staurolite Twins.

**Physical Properties.** The hardness is 7–7½, and the specific gravity is 3.65–3.75. A poor cleavage is present on some crystals parallel to the side pinacoid. The luster is vitreous, and the color is reddish, yellowish brown, brownish black, or gray.

**Composition.** Staurolite is a silicate of iron and aluminum, $FeAl_5Si_2O_{12}(OH)$, but is often impure. It is infusible and insoluble.

**Occurrence.** Staurolite occurs frequently with garnet, tourmaline, kyanite, or sillimanite in mica schist and gneiss. It is found in many localities in New England, and very perfect crystals are found in Fannin County, Georgia, and in North Carolina.

## SPHENE, $CaTiSiO_5$

Sphene, formerly called *titanite* in the United States, takes its name from the Greek word meaning wedge, in allusion to the characteristic type of crystal. Clear crystals of sphene can be cut into lovely gems, but such high quality is rare. In certain places in the world, as on the Kola Peninsula in Russia, where sphene is abundant, it is mined as an ore of titanium.

**Habit.** Sphene is monoclinic and frequently in well-formed wedge-shaped crystals as shown in Figs. 386 and 387. It may also be lamellar with a well-defined parting plane parallel to the pyramid.

**Physical Properties.** There is prismatic cleavage as well as the parting mentioned above. The hardness is 5–5½; the specific gravity is 3.4–3.55. The color is usually various shades of yellow to brown but may also be gray, green, or black. The luster is resinous to adamantine.

**Composition.** Sphene is calcium titanosilicate, $CaTiSiO_5$. It fuses with difficulty (4 in the scale) and is only slightly attacked by hydrochloric acid.

Fig. 386.

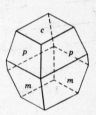

Fig. 387.

Sphene.

**Occurrence.** Sphene is one of the common accessory minerals in igneous rocks and as such is found in tiny crystals disseminated through granites, syenites, and nepheline syenites. It is found in larger crystals in metamorphic rocks and crystalline limestone.

As mentioned above, sphene is found on the Kola Peninsula, Russia, in large quantities. Here it is mined as a granular aggregate of crystals associated with nepheline. Fine crystals have come from St. Gothard, Switzerland; Zillertal, Tyrol; Arendal, Norway. In the United States crystals have been found at Riverside, California, and at several localities in New York.

# CHAPTER VIII

## On the Determination
## of Minerals

To the beginner it will seem a difficult thing to become so familiar with the many different mineral species as to be able to recognize each of them at sight; it is difficult, in fact impossible, even for the trained mineralogist to be always prompt and sure in his determination. For there are a large number of distinct species, between 1,500 and 2,000, many of them very rare, and not a few appear in a great number of varieties. The varieties sometimes depend upon fundamental differences of chemical composition, as among the garnets; and sometimes upon less essential distinctions of habit, color, and state of aggregation, as with the varieties of quartz, calcite, and fluorite. Hence it is obvious that the properties that can be seen at once, without aid of careful tests, are often insufficient to determine a mineral positively. The experienced mineralogist, although he learns to know minerals so well that he can name most of them at sight, is ever distrustful of himself and often hesitates to give the name quartz to a specimen having the appearance of this common species without, for example, a confirmatory test of hardness. Confidence and hasty judgment belong to those who have little experience and a scanty knowledge of the difficulties of the subject.

On the other hand, to recognize most of the minerals that are likely to be collected on a mineralogical excursion or to be obtained by exchange with other collectors is generally easy even for the beginner. For the number of common species is small, and quartz, feldspar, mica, calcite, barite, and, among the metallic species, galena, sphalerite, pyrite, and chalcopyrite are constantly presenting themselves. Although their properties vary somewhat widely in different specimens, they are usually distinct, and in most cases a simple test will suffice for identification.

First of all, then, the mineralogist should know the common species well, for the chances are many times greater that an unknown specimen is one of them rather than a rare and little-known species. It may be rare, even a new one not before described and not given in any of the books; but this is a chance that does not often happen. A real difficulty, that even much experience does not entirely remove, lies in the fact that at any large mineral locality there are likely to be many nondescript specimens that show few distinct properties. These may be mixtures of several species, and often they arise from chemical decomposition of well-known minerals. About such specimens it may perhaps be impossible to say anything definite; in fact, exhaustive microscopic, x-ray, and chemical work are often needed to prove their identity. In such cases the beginner may well turn to someone more experienced for counsel.

The best way, then, is for one with a specimen of an unknown mineral in hand to think of the common species first and of others that may suggest themselves afterward, running over in mind or by reference to the book the properties observed and those of the species to which it is provisionally referred. Care should be exercised not to decide too hastily, but to give each property full weight. Do not give the name albite to a specimen of barite because it is in tabular glassy crystals, and overlook the fact that it is much too heavy as well as too soft. Do not give the name beryl to a crystal of apatite because it is a green hexagonal prism, and overlook the fact that it is much too soft. Finally, do not hesitate to confess ignorance—this the experienced mineralogist is ever ready to do, and it is this fact that enables him from time to time to identify some rare and interesting species and perhaps occasionally one new to science.

In the systematic determination of an unknown specimen the first step, as was insisted upon in Chapter II, is to learn all that it is possible to learn about it by looking at and handling it. It has already been shown that in this way its habit may be at least partially determined; also its cleavage, if it shows any; and finally its luster, color, degree of transparency, etc. But at the same time the other senses must be kept on the alert, so that, for example, if the specimen is particularly heavy or light, greasy to the touch, etc., all these points will be quickly perceived and duly regarded.

Then a touch with the point of a knife blade will show something as to the hardness. This, it must be repeated, should be done carefully so as not to injure the specimen. If the mineral is not scratched by a knife, it will be well to see whether it is scratched by or will scratch the smooth surface of a quartz crystal.

At the same time as the test for hardness, the streak, or color of the powder as seen best on a surface of ground glass or unglazed porcelain, must be noticed. Also, if the blade of the knife is magnetized, the distinguishing character of magnetite and pyrrhotite will show itself at once. The careful determination of the specific gravity requires more time and may be postponed until the blowpipe has been used, but a rough estimate of weight should already have been made by the hand.

Further, when all the properties mentioned have been noted, it will often be necessary to make some chemical tests (read carefully pp. 117 to 119). A fragment for examination can generally be obtained without injury to the specimen by a careful blow with a light hammer. This, powdered and placed on a watch glass with a little strong hydrochloric (or nitric) acid (p. 118), will effervesce with a nearly odorless gas ($CO_2$) if it is a carbonate.

The solution obtained will give the chemist the means of learning more (e.g., the presence of copper, p. 118), and if the specimen is insoluble, even when finely powdered and heated in acid, that is also an important bit of information.

The blowpipe tests may come before or after the other chemical examination, and these have been so fully explained in Chapter VI that they need not be repeated here. A careful study of this chapter will have given the student full command of this part of the subject, and his experience should have taught him what order is best for the different tests. He will have learned, for example, that a mineral with metallic luster should be tried first in the open tube. If sulfur is present (the mineral being a sulfide), it will be given off as sulfur dioxide (pp. 115, 116), and at the same time arsenic, antimony, and mercury will show themselves (p. 116). The closed-tube test may be made next and then the charcoal (p. 113), which will confirm the results already obtained and also show by the coating the presence of zinc, lead, etc.; a magnetic residue will indicate iron. Further, after roasting off (p. 110) the sulfur, arsenic, or antimony, the residue may be tested for copper, cobalt, etc., with borax on the platinum wire.

A mineral with nonmetallic luster may be tried first in the forceps, and the degree of fusibility, the flame coloration, and other phenomena noted (p. 106). Then an examination should be made on the platinum wire in the borax bead.

Note, finally, that to obtain correct and concordant results the pure mineral must be experimented upon. In many specimens two or more species are so closely mixed that it needs sharp eyes, aided by a magnifying glass,* to separate them; this is particularly true of metallic minerals. Many species commonly occur in an earthy mass so that it is difficult to obtain absolutely pure material. In that event the quartz or clay will often do no harm if its presence is noted and the results interpreted correctly. A fragment of cinnabar is entirely volatile on charcoal or in the tube, but frequently it is associated with a gangue of clay, and then this will of course be left behind. Such a fragment heated in the glass tube often yields water which comes from the nonessential gangue.

Even if at the beginning it seemed as if very little were known about a specimen, the careful use of the eyes, the hand, and the various tests which may be made in a few minutes will give a pretty complete table of its properties, and these may be used to fill out the blank list as suggested on p. 121. Usually, unless the specimen is very rare and unusual, it will be possible to suggest the name of a species with the description of which it is to be compared. Where this method of attack yields no definite result, the determinative tables on the following pages may be employed.

It should be pointed out that one should not rely completely on determinative tables, for in using them there is a tendency merely to make the observation and look it up in the table, rather than to sum up mentally the properties which should indicate a given species.

## DETERMINATIVE TABLES

In the following tables the minerals are grouped according to their physical properties; quickly made physical tests on an unknown specimen give the necessary data to use them.

The first division is on the basis of luster: (1) *metallic and sub-*

---

* Every mineralogist should have a pocket magnifying glass, for even good eyes often need assistance, especially in examining small crystals.

*metallic,* (2) *nonmetallic.* In the first group are placed those minerals that have the appearance of a metal and are quite opaque even on their thin edges. Most of these minerals have a colored streak that may be quite different from the color of the mineral, and thus this property is used in grouping the metallic minerals. The second group includes all those minerals with a nonmetallic luster; those that will transmit light on their thin edges. These in general give a colorless or light-colored streak which in only a few cases is diagnostic. Therefore, streak is not used in the grouping of the nonmetallic minerals.

The tables are next subdivided according to hardness. For metallic minerals the divisions are: (1) less than $2\frac{1}{2}$; (2) greater than $2\frac{1}{2}$, less than $5\frac{1}{2}$; (3) greater than $5\frac{1}{2}$. For nonmetallic minerals the divisions are: (1) less than $2\frac{1}{2}$; (2) greater than $2\frac{1}{2}$, less than 3; (3) greater than 3, less than $5\frac{1}{2}$; (4) greater than $5\frac{1}{2}$, less than 7; (5) greater than 7. These limits are used because it is easy to determine them. For example, $2\frac{1}{2}$ is the hardness of the fingernail, and minerals scratched by the fingernail are in the first division—under $2\frac{1}{2}$. A copper coin has a hardness of 3; the knife blade, $5\frac{1}{2}$; and a quartz crystal, 7. Thus without special equipment it is possible to determine in which subdivision to look for the name of the unknown specimen. As stated earlier in this book, caution should be used in making the hardness test. Make sure that the scratch made by the knife blade is truly a scratch and is not merely steel that has been taken off by a hard mineral. It is well to try the scratch both ways; that is, try to scratch the knife with the mineral after scratching the mineral with the knife. Finally, it should be pointed out that the state of aggregation may influence the hardness determination. For example, crystals of hematite have a hardness of 6, but when the mineral is compact and in a pulverulent aggregate it has a hardness of only 1–2. Minerals of this type have been listed under both hardness groupings in the tables.

For the nonmetallic minerals a further subdivision is made on the basis of cleavage. It will be noted that *prominent* cleavage is specified; one that the student should be able to perceive easily. However, a finely aggregated mineral may show no cleavage even though coarse crystals have perfect cleavage. When such a condition is common the mineral is listed under both headings.

It should be remembered that the tables are far from complete, although all the common minerals are listed.  Thus it is possible, although unlikely, that the unknown specimen may be a rare mineral that is not listed.  Further, since only the physical properties are used, it is quite impossible by use of the tables alone to identify all the minerals.  Sometimes one will find that the data he has apply to two or three species.  These minerals should then be looked up in the chapter on *Description of the Mineral Species*, and a blowpipe or other tests should be made to determine them definitely.

In order to make the tables as compact as possible, the following abbreviations are used:  H. = hardness, G. = specific gravity.

On page 291 is given a general outline of the classification. After determining the luster, the hardness, and the presence or absence of cleavage, one should consult this outline to find on which page in the tables the name of the unknown specimen appears.

# OUTLINE OF DETERMINATIVE TABLES

## METALLIC OR SUBMETALLIC LUSTER

1. Hardness less than 2½. Can be scratched by the fingernail and in general will leave a mark on paper. Page 292.

2. Hardness greater than 2½, less than 5½. Can be scratched by the knife blade, but cannot be scratched by the fingernail. Page 293.

3. Hardness greater than 5½. Cannot be scratched by the knife blade. Page 295.

## NONMETALLIC LUSTER

1. Hardness less than 2½. Can be scratched by the fingernail.
   a. Prominent cleavage. Page 296.
   b. No prominent cleavage. Page 296.

2. Hardness greater than 2½, less than 3. Cannot be scratched by the fingernail but can be scratched by a copper coin.
   a. Prominent cleavage. Page 297.
   b. No prominent cleavage. Page 298.

3. Hardness greater than 3, less than 5½. Cannot be scratched by a copper coin but can be scratched by the knife blade.
   a. Prominent cleavage. Page 299.
   b. No prominent cleavage. Page 301.

4. Hardness greater than 5½, less than 7. Cannot be scratched by the knife blade but can be scratched by quartz.
   a. Prominent cleavage. Page 303.
   b. No prominent cleavage. Page 304.

5. Hardness greater than 7. Cannot be scratched by quartz.
   a. Prominent cleavage. Page 306.
   b. No prominent cleavage. Page 306.

## METALLIC OR SUBMETALLIC LUSTER
HARDNESS LESS THAN $2\frac{1}{2}$
(Will leave a mark on paper)

| Color | Streak | H. | G. | Remarks | Name |
|---|---|---|---|---|---|
| Black | Black | 1–2 | 4.7 | Fibrous aggregates | Pyrolusite p. 174 |
| Black | Black | $1$–$1\frac{1}{2}$ | 2.3 | Greasy feel | Graphite p. 137 |
| Blue-black | Black | $1$–$1\frac{1}{2}$ | 4.7 | One good cleavage | Molybdenite p. 157 |
| Black | Gray-black | $2\frac{1}{2}$ | 7.6 | Good cleavage in 3 directions | Galena p. 141 |
| Blue-black | Gray-black | 2 | 4.5 | One good cleavage | Stibnite p. 151 |
| Red | Red | $2$–$2\frac{1}{2}$ | 8.1 | Adamantine luster | Cinnabar p. 149 |
| Red | Red-brown | 1 plus | 5.2 | Earthy | Hematite p. 165 |
| Yellow-brown | Yellow-brown | 1 plus | 3.6–4.0 | Earthy | Limonite p. 175 |
| Gray-black | Gray-black | $2$–$2\frac{1}{2}$ | 7.3 | Easily sectile | Argentite p. 139 |
| Blue | Black | $1\frac{1}{2}$–2 | 4.6 | Platy masses or coatings | Covellite p. 140 |

## METALLIC OR SUBMETALLIC LUSTER
### HARDNESS 2½ TO 5½
(Can be scratched by a knife)

| Color | Streak | H. | G. | Remarks | Name |
|---|---|---|---|---|---|
| Gray-black | Black | 3–4½ | 4.7–5.0 | Tetrahedral crystals or massive | Tetrahedrite p. 159 |
| Gray-black | Gray-black | 2½–3 | 5.7 | Imperfectly sectile | Chalcocite p. 140 |
| Steel-Gray | Black | 2–2½ | 7.3 | Easily sectile | Argentite p. 139 |
| Gray-black | Black | 3 | 4.4 | Usually shows cleavage | Enargite p. 160 |
| Gray to black | Black | 2½ | 7.5 | Good cleavage in 3 directions | Galena p. 141 |
| Tin-white | Gray-black | 3½ | 5.7 | Usually in fibrous masses | Arsenic p. 132 |
| Tin-white | Gray-black | 2 | 8.0–8.2 | Fuses easily, one cleavage | Sylvanite p. 158 |
| Brass-yellow Tin-white | Gray-black | 2½ | 9.4 | Fuses in candle flame | Calavarite p. 158 |
| Copper-red | Black | 5–5½ | 7.8 | Usually massive | Niccolite p. 148 |
| Brownish bronze | Black | 3 | 5.1 | Tarnishes purple | Bornite p. 141 |
| Bronze | Black | 4 | 4.6 | Magnetic | Pyrrhotite p. 147 |
| Brass-yellow | Black | 3½–4 | 4.1–4.3 | Sphenoidal crystals, commonly massive | Chalcopyrite p. 145 |
| Brass-yellow | Black | 3–3½ | 5.5 | Usually in radiating hairlike crystals | Millerite p. 148 |
| Black | Gray to black | 4 | 4.3 | In fibrous or crystalline masses | Manganite p. 177 |

## METALLIC OR SUBMETALLIC LUSTER
### Hardness 2½ to 5½ (*Continued*)
#### (Can be scratched by a knife)

| Color | Streak | H. | G. | Remarks | Name |
|-------|--------|-----|-----|---------|------|
| Black | Brown to black | 5½ | 4.6 | Pitchy luster | Chromite p. 169 |
| Brown to black | Brown to black | 5–5½ | 7.0–7.5 | One perfect cleavage | Wolframite p. 219 |
| Black | Black | 5–6 | 3.7–4.7 | Botryoidal, massive | Psilomelane p. 180 |
| Brown to black | Brown | 3½–4 | 3.9–4.1 | Dodecahedral cleavage; may be red | Sphalerite p. 144 |
| Black | Red-brown | 5½–6½ | 4.8–5.3 | Massive, reniform or micaceous | Hematite p. 165 |
| Red to black | Red | 2½ | 5.85 | Fuses in candle flame. Good cleavage. | Pyrargyrite p. 158 |
| Red | Red | 3½–4 | 6.0 | Massive or in isometric crystals | Cuprite p. 162 |
| Red | Red | 2–2½ | 5.55 | Fuses in candle flame. Good cleavage. | Proustite p. 158 |
| Brown to black | Yellow-brown | 5–5½ | 3.6–4.0 | Massive | Limonite p. 175 |
| Brown to black | Yellow-brown | 5–5½ | 4.37 | Radiating fibers, one cleavage | Goethite p. 175 |
| Red | Red | 2½ | 8.1 | Earthy, granular. May show cleavage. | Cinnabar p. 149 |
| Copper-red | Copper-red | 2½–3 | 8.9 | Malleable. Tarnishes black. | Copper p. 128 |
| Silver-white | Silver-white | 2½–3 | 10.5 | Malleable. Tarnishes black. | Silver p. 126 |
| Silver-white | Gray | 4–4½ | 14.–19. | Malleable | Platinum p. 130 |
| Silver-white | Silver-white | 2–2½ | 9.8 | Good cleavage, sectile, easily fusible | Bismuth p. 133 |
| Gold-yellow | Gold-yellow | 2½–3 | 15.0–19.3 | Malleable | Gold p. 124 |

# METALLIC OR SUBMETALLIC LUSTER
## HARDNESS GREATER THAN 5½
### (Cannot be scratched by a knife)

| Color | Streak | H. | G. | Remarks | Name |
|-------|--------|-----|-----|---------|------|
| Tin-white | Black | 5½–6 | 6.0–6.2 | Usually massive, granular | Arsenopyrite p. 156 |
| Tin-white | Black | 5½–6 | 6.1–6.9 | Usually massive | Smaltite p. 154 |
| Tin-white | Black | 5½ | 6.33 | Frequently in isometric crystals | Cobaltite p. 154 |
| Copper-red | Black | 5–5½ | 7.5 | Usually massive | Niccolite p. 148 |
| Brass-yellow | Black | 6–6½ | 5.0 | Massive and in isometric crystals | Pyrite p. 152 |
| Pale yellow | Black | 6–6½ | 4.9 | "Cock's comb" crystals, radiating masses common | Marcasite p. 155 |
| Black | Black | 6 | 5.18 | Strongly magnetic | Magnetite p. 168 |
| Black | Brown to black | 5½ | 9.0–9.7 | Pitchy luster. Massive granular. | Uraninite p. 222 |
| Black | Brown to black | 5½–6 | 4.7 | Platy crystals, massive; may be slightly magnetic | Ilmenite p. 166 |
| Black | Brown to black | 5–6 | 3.7–4.7 | Massive, stalactitic botryoidal | Psilomelane p. 180 |
| Brown to black | Dark brown | 5–5½ | 7.0–7.5 | One perfect cleavage | Wolframite p. 219 |
| Brown to black | Dark brown | 5½ | 4.6 | Pitchy luster | Chromite p. 169 |
| Black | Dark brown | 6 | 5.15 | Associated with zincite and willemite. Slightly magnetic. | Franklinite p. 169 |
| Black | Red-brown | 5½–6 | 4.8–5.3 | Radiating reniform. Rarely in crystals. | Hematite p. 165 |
| Brown to black | Yellow-brown | 5–5½ | 3.6–4.0 | Massive | Limonite p. 175 |
| Brown to black | Yellow-brown | 5–5½ | 4.37 | Radiating fibers, one cleavage | Goethite p. 175 |

*a. Shows good cleavage*

| Color | H. | G. | Remarks | Name |
|---|---|---|---|---|
| Brown, yellow, green, white | 2–2½ | 2.76–3.0 | Cleavage flakes elastic; tabular crystals | Muscovite p. 253 |
| Dark brown to black | 2½–3 | 2.95–3.0 | Cleavage flakes elastic; common black mica | Biotite p. 255 |
| Yellow-brown, green, white | 2½–3 | 2.86 | Cleavage flakes elastic. May show copperlike reflection. | Phlogopite p. 256 |
| Green | 2–2½ | 2.6–2.9 | Cleavage flakes flexible but not elastic | Chlorite p. 251 |
| White to dark green | 1 | 2.7–2.8 | Greasy feel. In foliated masses. | Talc p. 247 |
| White, gray, green | 2½ | 2.39 | Sectile. Pearly luster on cleavage. Flexible but not elastic. | Brucite p. 180 |
| Colorless or white | 2 | 1.99 | Soluble in water; bitter taste | Sylvite p. 182 |
| Colorless or white | 2 | 2.32 | In crystals, fibrous or massive without cleavage | Gypsum p. 217 |
| White | 2–2½ | 2.6 | Claylike. Will adhere to the dry tongue. | Kaolinite p. 246 |
| Lemon-yellow | 1½–2 | 3.49 | Resinous luster. Pale yellow streak. | Orpiment p. 150 |

*b. Shows no prominent cleavage*

| Color | H. | G. | Remarks | Name |
|---|---|---|---|---|
| Pearl-gray or colorless | 2–3 | 5.5 | Perfectly sectile. Turns brown on exposure to light. | Cerargyrite p. 183 |
| Pale yellow | 1½–2½ | 2.05–2.09 | Burns with a blue flame | Sulfur p. 133 |
| Yellow, brown, gray, white | 1–3 | 2.0–2.5 | Earthy, may be in rounded grains | Bauxite p. 178 |
| White | 1 | 1.95 | Usually in rounded masses, "cotton balls" | Ulexite p. 205 |
| Green to white | 2–3 | 2.2–2.8 | Usually in crusts and earthy masses | Garnierite p. 250 |
| Deep red to Orange | 1½–2 | 3.48 | Earthy. Fusible in candle flame. | Realgar p. 150 |

HARDNESS GREATER THAN 2½, LESS THAN 3
(Can be scratched by a copper coin)

*a. Shows good cleavage*

| Color | H. | G. | Remarks | Name |
|---|---|---|---|---|
| Lilac, gray, white | 2½–4 | 2.8–3.0 | Usually in small scales and irregular sheets. One cleavage. | Lepidolite p. 257 |
| Pink, gray, white | 3½–5 | 3.0–3.1 | Associated with emery. One cleavage. | Margarite p. 252 |
| Colorless or white | 3½ | 4.3 | Effervesces in cold hydrochloric acid. One cleavage. | Witherite p. 199 |
| Colorless or white | 3 | 1.95 | Two cleavages yielding splintery fragments | Kernite p. 205 |
| Colorless, white Rarely red, blue | 2½ | 2.1–2.3 | 3 cleavages at right angles. Common salt. | Halite p. 181 |
| Colorless, white | 2 | 1.99 | 3 cleavages at right angles. Bitter taste. | Sylvite p. 182 |
| Colorless, white, blue, gray, red | 3–3½ | 2.89–2.98 | 3 cleavages at right angles. Usually massive. | Anhydrite p. 216 |
| Colorless, white, variously tinted | 3 | 2.72 | 3 cleavages not at right angles. Effervesces in cold acid. | Calcite p. 187 |
| Colorless, white, pink | 3½–4 | 2.85 | Curved crystals, massive. 3 cleavages not at right angles. | Dolomite p. 192 |
| Colorless, white, blue, yellow, red | 3–3½ | 4.5 | Cleavage in 3 directions. Pearly luster on base. | Barite p. 212 |
| Colorless, white, blue, red | 3–3½ | 3.96 | Cleavage in 3 directions. Gives crimson flame of strontium. | Celestite p. 215 |
| Colorless, white, when impure, gray and brown | 3 | 6.2–6.4 | Adamantine luster. Massive or tabular crystals. | Anglesite p. 216 |

## NONMETALLIC LUSTER
### HARDNESS GREATER THAN 2½, LESS THAN 3 (*Continued*)
#### b. *Shows no prominent cleavage*

| Color | H. | G. | Remarks | Name |
|---|---|---|---|---|
| Colorless or white | 2–2½ | 1.7 | Soluble in water. Fuses in candle flame. | Borax p. 204 |
| Colorless or white | 2½ | 2.95–3.0 | Massive. May show pseudocubic parting. | Cryolite p. 183 |
| Colorless, white. Brown when impure | 3–3½ | 6.55 | Adamantine luster. Effervesces in nitric acid. | Cerussite p. 200 |
| Colorless or white | 3½ | 4.3 | Massive or in radiating masses. Effervesces in acid. | Witherite p. 199 |
| Colorless or white | 2–2½ | 2.6 | Compact, earthy | Kaolinite p. 246 |
| Yellow, brown, gray, white | 1–3 | 2.0–2.55 | Rounded grain and earthy masses | Bauxite p. 178 |
| Yellow, green, white, brown | 3½–4 | 2.33 | Usually in radiating or globular aggregates | Wavellite p. 211 |
| Dark green, yellow, green, white | 2–5 | 2.2 | Massive usually mottled. Fibrous in chrysolite asbestos. | Serpentine p. 248 |
| Green to turquois-blue | 2–4 | 2.0–2.4 | Massive compact. Gives test for copper. | Chrysocolla p. 265 |

HARDNESS GREATER THAN 3, LESS THAN $5\frac{1}{2}$

*a. Shows good cleavage*

| Color | H. | G. | Remarks | Name |
|-------|-----|-----|---------|------|
| Blue, white, gray, green | 5–7 | 3.56–3.66 | Bladed aggregates. H. = 5 parallel to length. H. = 7 across. | Kyanite p. 281 |
| White, yellow, brown, red | $3\frac{1}{2}$–4 | 2.1–2.2 | In sheaflike aggregates. Cleavage parallel to length. | Stilbite p. 243 |
| Colorless, white, green, yellow | $4\frac{1}{2}$–5 | 2.3–2.4 | Pearly luster on cleavage. Pseudocubic crystals. | Apophyllite p. 250 |
| White, yellow, red | $3\frac{1}{2}$–4 | 2.18–2.20 | Pearly luster on cleavage. Crystals tabular. | Heulandite p. 242 |
| Colorless, white | 4–$4\frac{1}{2}$ | 2.42 | One perfect cleavage. Decrepitates in candle flame. | Colemanite p. 206 |
| Colorless, white | $3\frac{1}{2}$ | 4.3 | Effervesces in cold acid. Gives green flame. | Witherite p. 199 |
| Colorless, white | $3\frac{1}{2}$–4 | 2.95 | Effervesces in cold acid. Pseudohexagonal twins. | Aragonite p. 197 |
| Colorless, white, gray | 5–$5\frac{1}{2}$ | 2.8–2.9 | Usually in cleavable masses | Wollastonite p. 264 |
| Colorless, white | 5–$5\frac{1}{2}$ | 2.25 | Slender prismatic crystals; radiating groups | Natrolite p. 244 |
| Colorless, white | $3\frac{1}{2}$–4 | 3.7 | Effervesces in cold acid. Prismatic crystals. | Strontianite p. 200 |
| White, pale green, blue | $4\frac{1}{2}$–5 | 3.4–3.5 | In radiating groups rarely showing cleavage | Hemimorphite p. 269 |
| White, green, black | 5–6 | 3.0–3.3 | Crystals usually slender. 2 cleavage directions at 55°. | Amphibole group p. 257 |
| White, green, black | 5–6 | 3.1–3.5 | Crystals usually short and stout. 2 cleavages at nearly 90°. | Pyroxene group p. 260 |
| Rose-red, pink-brown | $5\frac{1}{2}$–6 | 3.58–3.70 | Usually massive; cleavable | Rhodonite p. 263 |

*a. Shows good cleavage (Continued)*

| Color | H. | G. | Remarks | Name |
|---|---|---|---|---|
| Colorless, white, variously tinted | 3 | 2.72 | Effervesces in cold acid. 3 cleavage directions not at right angles. | Calcite p. 187 |
| Colorless, white, pink | 3½–4 | 2.85 | Crystals curved rhombohedrons. Pearly luster. | Dolomite p. 192 |
| White, yellow, gray, brown | 3½–5 | 3.0–3.2 | In dense compact masses or cleavable aggregates | Magnesite p. 193 |
| Light to dark brown | 3½–4 | 3.83–3.88 | Cleavable masses. Magnetic after heating. | Siderite p. 195 |
| Pink, red, brown | 3½–4½ | 3.45–3.60 | In cleavable masses | Rhodocrosite p. 195 |
| Brown, green, blue, white | 5 | 4.35–4.49 | Effervesces in cold acid. In botryoidal aggregates. | Smithsonite p. 196 |
| White, yellow, red | 4.5 | 2.05–2.15 | In rhombohedral crystals | Chabazite p. 244 |
| Colorless, white, blue, gray, red | 3–3½ | 2.89–2.98 | Usually in fine aggregates without cleavage | Anhydrite p. 216 |
| Colorless, white, blue, yellow, red | 3–3½ | 4.5 | Frequently in platy aggregates. Pearly luster. 3 cleavages. | Barite p. 212 |
| Colorless, white, red, blue | 3–3½ | 3.95 | Gives crimson flame. 3 cleavages. | Celestite p. 215 |
| Colorless, violet, green, yellow, brown | 4 | 3.18 | Cubic crystals with octahedral cleavage. | Fluorite p. 184 |
| White, pink, gray, green, brown | 5–6 | 2.65–2.74 | Prismatic crystals. 4 poor cleavage directions. | Scapolite p. 240 |
| Yellow, brown, black, red | 3½–4 | 3.9–4.1 | Resinous luster. Dodecahedral cleavage. | Sphalerite p. 144 |
| Blue, white, gray, green | 5½–6 | 2.15–2.3 | Massive or embedded grains in rock. | Sodalite p. 239 |

## NONMETALLIC LUSTER
### Hardness Greater Than 3, Less Than 5½ (Continued)
*b. Shows no prominent cleavage*

| Color | H. | G. | Remarks | Name |
|-------|-----|-----|---------|------|
| Colorless, pale green, yellow | 5–5½ | 2.8–3.0 | Usually in brilliant crystals | Datolite p. 277 |
| White, green, blue | 4½–5 | 3.4–3.5 | Usually in radiating crystal groups | Hemimorphite p. 269 |
| White, pink, gray, green | 5–6 | 2.65–2.74 | In prismatic crystals; granular or massive | Scapolite p. 240 |
| Colorless, white | 3½–4 | 2.95 | Effervesces in cold acid. Radiating crystals. | Aragonite p. 197 |
| Colorless, white | 5–5½ | 2.27 | Usually in trapezohedrons embedded in rock | Analcime p. 245 |
| Colorless, white | 3½–4 | 3.7 | Effervesces in cold acid. Prismatic crystals, massive. | Strontianite p. 200 |
| Colorless, white | 3½–5 | 3.0–3.2 | Commonly in dense compact masses | Magnesite p. 193 |
| Colorless, white | 3½ | 4.3 | Effervesces in cold acid. In radiating masses. | Witherite p. 199 |
| Colorless, white | 5–5½ | 2.25 | In slender prismatic crystals in rock cavities | Natrolite p. 244 |
| White, grayish red | 4 | 2.6–2.8 | Usually massive | Alunite p. 117 |
| May be any color | 5–6 | 1.9–2.2 | Conchoidal fracture | Opal p. 231 |
| Brown, green, blue, white | 5 | 4.35–4.40 | Effervesces in cold acid. In botryoidal masses. | Smithsonite p. 196 |
| Brown, gray, green, yellow | 5–5½ | 3.4–3.55 | Adamantine to resinous luster. Wedge-shaped crystals. | Sphene p. 284 |
| Colorless, white, yellow, red, brown | 3 | 2.72 | Effervesces in cold acid. May be fibrous, fine granular. | Calcite p. 187 |

## NONMETALLIC LUSTER
### HARDNESS GREATER THAN 3, LESS THAN 5½
*b. Shows no prominent cleavage (Continued)*

| Color | H. | G. | Remarks | Name |
|---|---|---|---|---|
| Light to dark brown | 3½–4 | 3.83– 3.88 | In compact concretions. Becomes magnetic on heating. | Siderite p. 195 |
| White, yellow, green, brown | 4½–5 | 5.9– 6.1 | Adamantine to vitreous luster. Usually in square tabular crystals. Fluorescent. | Scheelite p. 220 |
| Yellow, orange, red, green | 3 | 6.8 | Adamantine luster. In square tabular crystals. | Wulfenite p. 221 |
| Yellow, brown, gray, white | 1–3 | 2.0– 2.55 | In rounded grains and earthy masses | Bauxite p. 178 |
| Green, blue, brown, violet | 5 | 3.15– 3.20 | In hexagonal prisms, also massive | Apatite p. 207 |
| Green, brown, yellow, gray | 3½–4 | 6.5– 7.1 | In hexagonal crystals often cavernous | Pyromorphite p. 208 |
| Yellow, green, white, brown | 3½–4 | 2.33 | Characteristically in radiating aggregates | Wavellite p. 211 |
| Blackish green, yellow-green, white | 2–5 | 2.2 | Massive variety mottled green. Fibrous variety asbestos. | Serpentine p. 248 |
| Green, white, gray, brown | 5½ | 3.9– 4.2 | Massive, associated with zincite and franklinite. Fluorescent. | Willemite p. 271 |
| White, gray, blue, green | 5½–6 | 2.15– 2.3 | Massive or in embedded grain in rock | Sodalite p. 239 |
| Azure-blue, greenish blue | 5–5½ | 2.4– 2.45 | Massive, associated with pyrite | Lazurite p. 240 |
| Bright green | 3½–4 | 3.9– 4.0 | Mammillary, fibrous, associated with azurite. Effervesces in cold acid. | Malachite p. 201 |
| Blue | 3½–4 | 3.77 | In small crystals with malachite. Effervesces in cold acid. | Azurite p. 203 |

| Color | H. | G. | Remarks | Name |
|---|---|---|---|---|
| White, gray, lavender, green | 6½–7 | 3.35–3.45 | In thin tabular crystals. Associated with emery. | Diaspore p. 175 |
| Grayish green, brown | 6–7 | 3.23 | In long slender crystals in schists | Sillimanite p. 281 |
| Yellowish to green-black | 6–7 | 3.35–3.45 | Cleavage parallel to length of crystals | Epidote p. 275 |
| Blue, gray, green, white | 5–7 | 3.56–3.66 | Bladed aggregates. H. = 5 parallel to length. H. = 7 across. | Kyanite p. 281 |
| White, green, blue | 6 | 3.0–3.1 | In cleavable masses resembling feldspar | Amblygonite p. 211 |
| Colorless, white, gray | 5–5½ | 2.8–2.9 | Usually in cleavable masses | Wollastonite p. 264 |
| Colorless, white | 5–5½ | 2.25 | In slender prismatic crystals in rock cavities | Natrolite p. 244 |
| White, gray, red, green | 6 | 2.54–2.56 | In cleavable masses and irregular grains in rocks | Orthoclase Microcline p. 233 |
| Colorless, white, gray | 6 | 2.62–2.76 | In cleavable masses and irregular grains in rocks. Show striations on one cleavage. | Plagioclase p. 236 |
| White, gray, pink, green | 6½–7 | 3.15–3.20 | In flattened striated crystals. Found in pegmatites. 2 cleavages at nearly 90°. | Spodumene p. 262 |
| White, green, black | 5–6 | 3.1–3.5 | In stout prisms and irregular grains in rock. 2 cleavages at nearly 90°. | Pyroxene group p. 260 |
| Brown, green, bronze, black | 5½ | 3.2–3.5 | Crystals usually prismatic. Fibrous and massive; 2 cleavages at nearly 90°. | Enstatite p. 260 |
| Red, pink, brown | 5½–6 | 3.58–3.70 | Usually massive, cleavable to compact; 2 cleavages at nearly 90°. | Rhodonite p. 263 |
| White, green, black | 5–6 | 3.0–3.3 | Crystals slender may be fibrous—asbestos; 2 cleavages at 55°. | Amphibole group p. 257 |
| Blue, gray, white, green | 5½–6 | 2.15–2.3 | Massive or in embedded grains; 6 cleavage directions. | Sodalite p. 239 |

| Color | H. | G. | Remarks | Name |
|---|---|---|---|---|
| Colorless, white | 5–5½ | 2.27 | In trapezohedral crystals in rock cavities | Analcime p. 245 |
| May be any color | 5–6 | 1.9– 2.2 | Conchoidal fracture. May show play of colors. | Opal p. 231 |
| Colorless, green, yellow | 5–5½ | 2.8– 3.0 | Usually in brilliant crystals in rock cavities | Datolite p. 277 |
| Gray, white, colorless | 5½–6 | 2.45– 2.50 | In trapezohedral crystals embedded in dark rock | Leucite p. 238 |
| White, pink, gray, green | 5–6 | 2.65– 2.74 | In prismatic crystals, granular massive | Scapolite p. 240 |
| Colorless, white, variously colored | 7 | 2.65 | In crystals with horizontal striations on prisms. Massive. | Quartz p. 223 |
| White, gray, greenish, red | 5½–6 | 2.55– 2.65 | Greasy luster. A rock mineral, usually massive. | Nepheline p. 239 |
| Yellow, brown, orange | 6–6½ | 3.1– 3.2 | Occurs in grains in crystalline limestone. | Chondrodite p. 271 |
| Brown, yellow, red, green | 7 | 2.65 | Waxy luster. May be lining cavities. | Chalcedony p. 227 |
| Blue, bluish green, green | 6 | 2.6– 2.8 | Usually found in reniform and stalactitic masses. | Turquois p. 212 |
| Green, gray, white | 6–6½ | 2.8– 2.95 | Reniform and stalactitic with crystalline surface | Prehnite p. 252 |
| Green, white, gray, brown | 5½ | 3.9– 4.2 | Massive and in disseminated grains. Fluoresces. | Willemite p. 271 |
| Green, brown | 6½–7 | 3.27– 3.37 | Usually in disseminated grains in dark igneous rocks | Olivine p. 269 |
| Black and variously colored | 7–7½ | 3.0– 3.25 | In crystals with triangular cross section. Single crystals may show several colors. | Tourmaline p. 265 |
| Green, brown, yellow, blue | 6½ | 3.35– 4.45 | In square prismatic crystals, columnar, massive | Idocrase p. 274 |
| Brown, gray, green, yellow | 6½–7 | 3.27– 3.35 | In wedge-shaped crystals | Axinite p. 280 |

## NONMETALLIC LUSTER
### HARDNESS GREATER THAN 5½, LESS THAN 7
*b. Shows no prominent cleavage (Continued)*

| Color | H. | G. | Remarks | Name |
|---|---|---|---|---|
| Brown to black | 7–7½ | 3.65–3.75 | In prismatic crystals and penetration twins found in schists | Staurolite p. 282 |
| Red, brown, green | 7½ | 3.16–3.20 | Square prismatic crystals. May be altered to mica. Black cross in cross section (chiastolite). | Andalusite p. 280 |
| Brown, gray, green, yellow | 5–5½ | 3.4–3.55 | Adamantine luster. Crystals wedge-shaped. | Sphene p. 284 |
| Brown to black | 6–7 | 6.8–7.1 | May have reniform surface. Prismatic crystals rare. | Cassiterite p. 171 |
| Reddish brown to black | 6–6½ | 4.18–2.25 | Prismatic crystals vertically striated | Rutile p. 173 |
| Blue, colorless | 7–7½ | 2.60–2.66 | In embedded grains resembling quartz | Cordierite p. 246 |
| Azure-blue, greenish blue | 5–5½ | 2.4–2.45 | Usually massive. Associated with pyrite. | Lazurite p. 240 |
| Blue, green, white, gray | 5½–6 | 2.15–2.3 | Massive or in embedded grains. Poor cleavage. | Sodalite p. 239 |

# NONMETALLIC LUSTER

*a. Shows good cleavage*

| Color | H. | G. | Remarks | Name |
|---|---|---|---|---|
| Colorless, yellow, pink, blue, green | 8 | 3.4–3.6 | One perfect cleavage. Usually in crystals. | Topaz p. 279 |
| Brown, gray, greenish gray | 6–7 | 3.23 | In long slender crystals. Found in schists. | Sillimanite p. 281 |
| White, gray, pink, green | 6½–7 | 3.15–3.20 | In flattened striated crystals. 2 cleavages at nearly 90°. | Spodumene p. 262 |
| Colorless, yellow, red, black | 10 | 3.5 | Adamantine luster. In octahedral crystals; 4 cleavage directions. | Diamond p. 134 |
| Colorless and almost any color | 9 | 4.0 | Adamantine luster. No cleavage but good parting at nearly 90°. | Corundum p. 163 |

*b. Shows no prominent cleavage*

| Color | H. | G. | Remarks | Name |
|---|---|---|---|---|
| Colorless, white, variously colored | 7 | 2.66 | Crystals show horizontal striations on prism faces. | Quartz p. 223 |
| Colorless and almost any color | 9 | 4.0 | Adamantine luster. Parting fragments nearly cubic. | Corundum p. 163 |
| Red, black, blue, green, brown | 8 | 3.6–4.0 | Commonly in octahedral crystals | Spinel p. 167 |
| Green, yellow, pink, colorless | 7½–8 | 2.75–2.8 | In hexagonal prisms. Poor basal cleavage. | Beryl p. 268 |
| Yellowish to emerald-green | 8½ | 3.65–3.8 | Frequently in tabular striated crystals | Chrysoberyl p. 171 |
| Black and variously colored | 7–7½ | 3.0–3.25 | In slender prismatic crystals with triangular cross section | Tourmaline p. 265 |
| Green, gray, white | 6½–7 | 3.3–3.5 | Massive, closely compact, difficult to break | Jadeite and Nephrite p. 258 |

## NONMETALLIC LUSTER

*b. No prominent cleavage (Continued)*

| Color | H. | G. | Remarks | Name |
|-------|----|----|---------|------|
| Green to brown | 6½–7 | 3.27–3.37 | Usually in disseminated grains in dark igneous rocks | Olivine p. 269 |
| Reddish brown to black | 6–7 | 6.8–7.1 | May have reniform surface. Prismatic crystals rare. | Cassiterite p. 171 |
| Brown, red, green | 7½ | 3.16–3.20 | Square prismatic crystals. Cross section may show black cross (chiastolite). | Andalusite p. 280 |
| Brown, green, yellow, gray | 6½–7 | 3.27–3.35 | In wedge-shaped crystals with sharp edges | Axinite p. 280 |
| Brown to black | 7–7½ | 3.65–3.75 | In prismatic crystals and cruciform twins. In schists. | Staurolite p. 282 |
| Brown, red, gray, colorless | 7½ | 4.68 | Usually in small prismatic crystals | Zircon p. 276 |
| Brown, red, yellow, green | 6½–7½ | 3.5–4.3 | Usually in dodecahedrons or trapezohedrons | Garnet p. 271 |

# APPENDIX I

## *List of Common Minerals Arranged According to Prominent Elements*

CARBON      Diamond.

Graphite.

HYDROGEN      Ice (and water).

ARSENIC      Native arsenic.

Realgar and orpiment, arsenic sulfides.

ANTIMONY      Native antimony.

Stibnite, antimony sulfide.

BISMUTH      Native bismuth.

MOLYBDENUM      Molybdenite, molybdenum sulfide.

Wulfenite, lead molybdate.

GOLD      Native gold.

Calaverite, gold telluride.

Sylvanite, gold-silver telluride.

PLATINUM      Native platinum.

SILVER      Native silver.

Argentite, silver sulfide.

Pyrargyrite, sulfide of silver and antimony.

Proustite, sulfide of silver and arsenic.

Cerargyrite, silver chloride.

MERCURY      Native mercury.

Cinnabar, mercury sulfide.

COPPER      Native copper.

Chalcocite, copper sulfide.

Bornite and chalcopyrite, sulfides of copper and iron.

Tetrahedrite, sulfide of antimony and copper.

Cuprite, cuprous oxide.

Malachite and azurite, carbonates of copper.

Dioptase and chrysocolla, silicates of copper.

LEAD      Galena, lead sulfide.

Jamesonite and bournonite, sulfides of antimony and lead.

Pyromorphite, lead phosphate.

Mimetite, lead arsenate.

Vanadinite, lead vanadate.

Cerussite, lead carbonate.

Anglesite, lead sulfate.
Crocoite, lead chromate.
Wulfenite, lead molybdate.

TIN     Cassiterite, tin dioxide.

TITANIUM     Rutile, octahedrite, and brookite, all titanium dioxide.
Ilmenite, ferrous titanate.

URANIUM     Uraninite.
Torbernite and autunite, uranium phosphates.

IRON     Native iron.
Pyrrhotite, iron sulfide.
Pyrite and marcasite, iron disulfides.
Arsenopyrite, iron sulfarsenide.
Hematite, iron sesquioxide.
Magnetite, magnetic iron oxide.
Franklinite, iron-zinc-manganese oxide.
Chromite, iron-chromium oxide.
Goethite and limonite, hydrated iron oxides.
Siderite, iron carbonate.
Also columbite, wolframite, and triphylite.

NICKEL     Millerite, nickel sulfide.
Niccolite, nickel arsenide.
Garnierite, nickel silicate.

COBALT     Linnaeite, cobalt sulfide.
Smaltite and cobaltite, arsenides of cobalt.
Erythrite, cobalt arsenate.

MANGANESE     Pyrolusite, manganite, and psilomelane, oxides of manganese.
Rhodonite, manganese silicate.
Rhodochrosite, manganese carbonate.

ZINC     Sphalerite, zinc sulfide.
Zincite, zinc oxide.
Franklinite, iron-zinc-manganese oxide.
Willemite and hemimorphite, zinc silicates.
Smithsonite, zinc carbonate.

ALUMINUM     Corundum, aluminum oxide.
Spinel, oxide of magnesium and aluminum.
Cryolite, fluoride of aluminum and sodium.
Turquois and wavellite, aluminum phosphates.
Amblygonite, phosphate of aluminum and lithium.

CALCIUM     Fluorite, calcium fluoride.
Calcite and aragonite, calcium carbonates.
Apatite, calcium phosphate.
Anhydrite, calcium sulfate.

Gypsum, hydrated calcium sulfate.
Scheelite, calcium tungstate.

MAGNESIUM   Brucite, magnesium hydroxide.
Magnesite, magnesium carbonate.
Boracite, magnesium borate.
Dolomite, calcium magnesium carbonate.

BARIUM   Barite, barium sulfate.
Witherite, barium carbonate.

STRONTIUM   Celestite, strontium sulfate.
Strontianite, strontium carbonate.

SODIUM and   Halite or rock salt, sodium chloride.
POTASSIUM   Borax, sodium borate.
Sylvite, potassium chloride.

SILICON   Quartz, silicon dioxide.
Opal, hydrated silicon dioxide.

The silicate minerals, too numerous to list here, are grouped together in this book under *silicates* in the last section in descriptive mineralogy.

# APPENDIX II

## *Most Important Minerals for a Small Collection*

The following list includes the names of the minerals that it is most important for the young mineralogist to have in his collection; they are printed in SMALL CAPITALS. To these are added, in ordinary type, a number of others which are also important, but less so; they may well be present in the cabinet of the school or academy.

GOLD in quartz
SILVER
COPPER
SULFUR
GRAPHITE
An ore of silver
Chalcocite
Bornite
GALENA
SPHALERITE
CHALCOPYRITE
PYRRHOTITE
Niccolite
MILLERITE
CINNABAR
Orpiment
STIBNITE
PYRITE
MARCASITE
ARSENOPYRITE
Molybdenite
Smaltite
TETRAHEDRITE
CUPRITE
Zincite
CORUNDUM
HEMATITE
Spinel
MAGNETITE

Franklinite
CHROMITE
CASSITERITE
RUTILE
Goethite
Manganite (or pyrolusite)
Brucite
HALITE
Cryolite
FLUORITE
CALCITE (several varieties)
DOLOMITE
Magnesite
SIDERITE
Rhodocrosite
SMITHSONITE
ARAGONITE
Strontianite
Witherite
CERUSSITE
MALACHITE
Azurite
APATITE
PYROMORPHITE
Amblygonite
Wavellite
BARITE
CELESTITE
Anglesite

Anhydrite
Gypsum
Wulfenite
Quartz (several varieties)
Opal
Orthoclase
Albite
Oligoclase
Labradorite
Nepheline
Scapolite
Heulandite
Stilbite
Chabazite
Natrolite
Analcime
Talc
Serpentine
Garnierite
Apophyllite
Chlorite
Prehnite

Muscovite
Biotite
Lepidolite
Amphibole (several varieties)
Pyroxene (several varieties)
Spodumene
Rhodonite
Tourmaline
Beryl
Hemimorphite
Olivine
Willemite
Garnet
Idocrase
Epidote
Zircon
Datolite
Topaz
Andalusite
Kyanite
Staurolite
Sphene

If the student limits himself to *small* specimens, as advised on p. 10, a collection including the species mentioned will not occupy a great deal of space and, if desired, can be purchased at no great cost. From time to time additional specimens can be obtained by exchange or purchase.

Of the minerals in the above list the following are most desirable for the blowpipe and other chemical tests described in Chapter VI. Suitable fragments, of the needed purity, can be purchased for a very small expenditure of money.

Stibnite, molybdenite, an ore of silver, cinnabar, chalcopyrite, tetrahedrite, cuprite or malachite, galena, pyromorphite, cassiterite, rutile, pyrite, arsenopyrite, hematite or siderite, millerite, rhodonite, sphalerite, corundum, cryolite, fluorite, calcite, apatite, brucite, barite, celestite, orthoclase, amphibole (actinolite), garnet (almandite), tourmaline, natrolite.

Also in addition to these: a mineral containing lithium, such as either spodumene or amblygonite; one containing cobalt, such as smaltite; one containing chromium, such as chromite; one containing vanadium, such as vanadinite; one containing uranium, such as uraninite (pitchblende) or autunite.

# Index

The words in **bold face** type are the names of the common minerals described in detail in the text.

Topaz, 279
 false, 227
Topazolite, 273
**Tourmaline, 265**
Translucent, 84
Transparency, 84
Trapezohedron, 24
Travertine, 191
**Tremolite, 258**
Triboluminescence, 87
Triclinic pinacoidal class, 44
Triclinic system, 44
**Tridymite, 230**
Troostite, 271
Tufa, 191
Tungstates, 219
Tungsten, ores of, 219, 220
 test for, 110
Turkey-fat ore, 196
**Turquois, 212**
Twin axis, 54
Twin crystals, 53
Twin plane, 54

**Ulexite, 205**
Uneven fracture, **72**
Uranophane, 223
**Uraninite, 222**
Uranium, 223
 test for, 110
**Uvarovite, 273**

Valence, 96
**Vanadinite, 210**
Vanadium, test for, 110
Variegated copper ore, 141
Verdantique, 248
Vermiculite, 255
Vesuvianite, 274
Vicinal plane, 51

Watch glass, 103
**Water, 161**
Water, tests for, 117
**Wavellite, 211**
Waxy luster, 82
Wernerite, 241
White iron pyrites, 155
**Willemite, 271**
**Witherite, 199**
**Wolframite, 219**
**Wollastonite, 264**
Wood opal, 231
Wood tin, 171
**Wulfenite, 221**

Zeolites, 241
Zinc, ores of, 144, 163
 test for, 113
Zinc blend, 144
**Zincite, 163**
**Zircon, 276**